Green Building
Handbook

Green Building

Handbook

Volume 1

A guide to building products and their impact on the environment

Tom Woolley,
Queens University of Belfast

Sam Kimmins, Paul Harrison and Rob Harrison
ECRA, Manchester

 Taylor & Francis
Taylor & Francis Group

LONDON AND NEW YORK

Green Building Digest

First published 1997 by Taylor & Francis

Reprinted 1998, 1999 by Taylor & Francis
Reprinted 2001 by Taylor & Francis
2 Park Square, Milton Park, Abingdon, Oxon, OX14 4RN
270 Madison Ave, New York NY 10016

Transferred to Digital Printing 2006

Taylor & Francis is an imprint of the Taylor & Francis Group, an informa business

Printed and bound in Great Britain by TJI Digital, Padstow, Cornwall

British Library Cataloguing in Publication Data
A catalogue record for this book is available from the British Library

Library of Congress Cataloguing in Publication Data
A catalogue record for this book is available from the Library of Congress

ISBN 10: 0-419-22690-7
ISBN 13: 978-0-419-22690-1

Publisher's note
This book has been produced from finished pages supplied on disk by the
authors.

Contents

Preface

This book is based on material from the first 12 issues of the Green Building Digest. We began to publish the digest in 1995 in an attempt to distribute information about the environmental impact of building materials and techniques to a wide range of people concerned with the built environment. It was partly financed by a grant from the Department of the Environment in England through its Environmental Action Fund, subscriptions, grants from other bodies, fund-raising and a great deal of voluntary effort. ACTAC, a federation of technical aid centres and consultancies, which has been involved in helping community groups taking environmental action for over 15 years, commissioned the Ethical Consumer Research Association to research material for a series of bi-monthly issues of digests on different materials and products. An advisory committee of ACTAC members and other experts in the green building field reviewed the drafts produced by ECRA and suggested topics for future issues of the digest.

Each of the chapters in the book is based on an issue which was distributed to subscribers. The number of subscribers to the digest has grown steadily as, over the last two to three years, interest in green building has increased enormously. Support for the digest is strong and it is hoped to continue producing the digest on a subscription basis, incorporating collected issues into further books. It was also the intention that past issues would be reviewed and updated as new information and research became available.

At the time of writing the digest was being relaunched through the Queens University, Belfast, and readers can obtain issues of the digest, not included here and information about further subscription from Queens University Belfast (01232 335 466) or ECRA *(details can be found at the back of the book)*. Those who have already seen the first 12 issues of the digest will find that there have been some small modifications in the transference to a book format. Some duplication has been omitted and the digests are now in a different and, hopefully, more logical order. We have

tried to bring the supplier information as up to date as possible, but inevitably such information changes as the green building field is expanding so rapidly at present.

Many people, including, in particular, the Association of Environment Conscious Builders, have been very supportive in the production of the digest and in the organisation of related events such as the annual **Green Buildings Fair** which has been organised annually in Leeds in 1995, '96 and '97. Indeed the whole project is an example of the spirit of sharing and co-operation which has long been the ethos of the technical aid movement since its foundation in the late 70s and early 80s. For this reason we are keen to get feedback from readers and for anyone with ideas and information to contribute, particularly with experience of applying green building principles and materials. Our objective is to circulate this experience and knowledge so that we can all benefit and thus, in the long run do something to mitigate the damage which is being done in the name of development and progress to the planet which we all inhabit. For this reason we have attempted to be transparent about the sources of our information and methods behind the production of the digest rather than present ourselves as *experts* , who restrict access to information and specialist technical knowledge only making it available on the payment of a substantial fee . Such expertise is an accumulation of shared knowledge available to all who want to protect the environment rather than something available to an elite.

Having said this, it is important to remember that to produce material of this quality and usefulness costs money and at present it is not easy to raise research funds in this field. Much of the accumulated knowledge is at a price and to continue this work, minor support through subscriptions to the Green Building Digest and more substantial funds will be needed. We are only scratching the surface of the problem and much more work needs to be done before we can feel confident that we know how to produce a perfect green building.

Tom Woolley Crossgar 1997

Acknowledgements

Ronnie Wright, Steve Smith, Lynn McCann, Debbie McCann, ACTAC Council of Management (1995 onwards), Edward Walker, Department of the Environment Environmental Action Fund, Charities Aid Foundation, Keith and Sally Hall (Association for Environment Conscious Builders (AECB), Miles Sibley, Polyp (Cartoons), Rita Harkin, Rachel Bevan

GBD Advisory Panel Members

Rob Bumbey (One Stop Architects), Christopher Day, Tom Smerdon (BSRIA), Jonathan Hines (architype), Rod Nelson (Soil Association), Heimir Salt, Bret Willers, Andrew Yates (Eco-Arc), Lindsay Halton, Steve Curwell, Sandy Haliday (Gaia Research), Tom Woolley, Keith Hall

Part 1

Introduction

Green Building 1

Plate 1. London Wildlife Trust Education Centre, Marsden Road, East Dulwich, South East London.

The body is a complex thing with many constituent parts, and to understand the behaviour of a whole living body you must apply the laws of physics to its parts not to the whole. ....We peel our way down the hierarchy until we reach units so simple that, for everyday purposes, we no longer feel the need to ask questions about them.

Richard Dawkins, The Blind Watchmaker (1)

1.1 The Nature of Green Building

There is a lot of general and rather superficial literature on green issues, much of it about social and economic policy or doom and gloom on the future of the planet. Many people who want to behave in an environmentally friendly way find such literature frustrating because it is often preaching to the converted. What they want to know, is not so much the general picture, though this is of course important, but more practical information on how to actually do things. They may come up against gurus who talk about holistic theories or a strange new esoteric language. Instead we have tried to create easy-to-manage packages which allow specifiers and clients to understand what is going on and to take responsible decisions about what to do. This is one of the main objects of the handbook. In this chapter, the background to this approach is explained through a brief review of the theories and basic principles of green building.

1.1.1 What is a Green Building?

It is necessary to explain the meaning of the term *Green Building* , why we are concerned with it and to set in context the writing of the digests which follow. We have to explain the methodology which underpins the assessment of products and materials and how *you* can decide whether something qualifies as green or not. In order to deal with these questions it is necessary to examine the philosophies which underlie environmental thinking and to warn the reader to come to his or her own conclusions about the issues raised rather than simply accepting that anyone has the final answers at this stage.

The relationship between this work and the fundamental principles of community technical aid and user participation in design are also examined because we firmly believe that genuine environmental action is only meaningful if it involves ordinary people taking charge of their environment at a local level.

Most people buying this book may already have concern for the environment. Though the word **Green** may have put some people off as it has political connotations, others deride it as a passing architectural stylistic fashion of buildings made out of unseasoned timber and grass roofs.

We assume that buildings are green if they if they look hand made and are built of natural materials........but working in aluminium and glass might in the long run create a more genuinely sustainable architecture.
Deyan Sudjic[2]

Sudjic's viewpoint comes from an attitude to architecture in which stylistic questions tend to be considered more important than environmental ones. But Sudjic alerts us to the danger of assuming that because a building looks superficially green it is creating less damage to the environment than one that looks 'high tech' or post modern.

For a building to be green it is essential for the environmental impact of all its constituent parts and design decisions to be evaluated. This is a much more thorough exercise than simply adding a few green elements such as a grass roof or a solar panel. The purpose of the digest is to help designers, specifiers and clients to make relatively objective decisions about the environmental impact of materials, products and building solutions with some reasonably hard facts, at least as far as the current state of the art (or science) permits.

Many people avoid the use of the term green altogether, especially those operating in a more commercial environment. They will talk about **environmentally friendly** buildings or **sustainable development.** Are these terms euphemisms or do they mean something different ? There is undoubtedly a need for some people to distance themselves from activists who climb up trees or dig tunnels in the path of new roads. There are many who fear that such associations will frighten off the relatively conservative construction industry which is just as involved in road construction as building houses or visitor centres.

Our approach is far less timid. The word green is unequivocal, it is a symbol of a desire to create a built environment which meets a whole range of criteria, without any fudging or attempts to soften the blow. Sadly there are those whose concern for the environment only extends to possible fresh marketing opportunities and it is not uncommon for companies to add environmental credentials to their advertising literature. So we don't apologise for talking about **Green Buildings.** On the other hand we have tried very hard to ensure that the information which has been digested is as objective as possible and relies on scientific and practical evidence, not ideological commitment. The aim has been to allow the **reader** to make ideological decisions rather than mixing this up with the practical data. Where there are questions about the issues being raised then this is made very explicit in the text.

We have also had to deal with the question of opposition from manufacturers and other vested interests in the current construction industry who might object to their product or material not being presented in the best light. Our approach has not been to launch attacks on companies, nor focus on the environmental crimes of particular companies, even though such activity is, sadly, all too common. This would have given the digest too negative an edge and is perhaps better done elsewhere. The negative environmental impacts of materials is a key issue in the digest and even this could be seen as being antagonistic to industry. In the case of *PVC*, the British Plastics Federation objected to our draft of the issue on rainwater goods and we decided to print their response in that chapter, so that readers can make up their own minds about both sides of the argument.

1.1.2 Defining Green Building

So how do others define green building?
Robert and Brenda Vale say:

" that a green approach to the built environment involves a holistic approach to the design of buildings ; that all the resources that go into a building, be they materials, fuels or the contribution of the users need to be considered if a sustainable architecture is to be produced."[3]

Stuart Johnson talks about

" how the environmental impact of individual properties can be mitigated."[4]

Sim Van Der Ryn and Stuart Cowan tell us we must

"infuse the design of .products, buildings and landscapes with a rich and detailed understanding of ecology."[5]

There are many such statements, too many to review here, but a comprehensive bibliography on the subject can be found at the end of the book. However, from a review of the literature the conclusion can be drawn that the words Green, Sustainable, Environmental, Ecological and so on are interchangeable. The nuances of their use depend on the context and the audience and thus the novice in the field will not get too much clear guidance from these labels. On the other hand it is important to be as clear as possible about the methodology employed to assess materials and products and methods of building and we cannot assume that everyone is talking about the same thing. There are undoubtedly many different shades of green!
In any case, general statements do not bring us much closer to a detailed understanding of how to create green buildings and as clients increasingly ask for their buildings to be green or environmentally friendly, professionals and construction industry bodies are having to wrestle with

these issues. For instance, the Building Services Research and Information Association (**BSRIA),** a mainstream construction industry body, defines **"sustainable construction"** as

"the creation and responsible management of a healthy built environment based on resource efficient and ecological principles."[6]

BSRIA tell us that these principles include:
* Minimising non-renewable resource consumption
* Enhancing the natural environment
* Eliminating or minimising the use of toxins
thus combining energy efficiency with the impact of materials on occupants.

Consultants, Sustainable Development Services in Seattle, USA, who provide a special consultancy service to clients, tell us that they provide analysis and integrated solutions in the following functional areas:

* Energy conservation
* Pollution prevention
* Resource efficiency
* Systems Integration
* Life Cycle Costing

They try to interpose themselves between clients and architects and builders to ensure that capital development proposals "reconcile the cultural, ecological and economic needs of society," before a brief or designs have been prepared.[7]

An examination of these statements makes it clear that producing green buildings involves resolving many conflicting issues and requirements. Each design decision, even the decision about what to build or where to build or even whether to build at all has environmental implications. Decisions about layout, relationship with site, the effects of wind and weather, possible use of solar energy, orientation, shading, ventilation, specification of materials and structural systems, must all be evaluated in terms of their impact on the environment **and** the occupants of buildings.

Green building is not simply about protecting the biosphere and natural resources from over-exploitation or over-consumption, nor is it simply about saving energy to reduce our heating bills, it considers the impact of buildings and materials on occupants and the impact of our lives on the future of the Earth.

1.2 Principles of Green Building

Because of the complexity of these issues it has been found useful to group consideration of green building under four headings . These are set out below with examples of the sorts of green building measures that can be taken under each of the headings:

(a) Reducing Energy in Use
for example
Use maximum possible low embodied energy insulation, but with good ventilation
Use low energy lighting and electrical appliances
Use efficient, low pollution heating
Make use of passive and active solar energy wherever feasible
Use passive and natural ventilation systems rather than mechanical

(c) Reducing Embodied Energy and Resource Depletion
for example
Use locally sourced materials
Use materials found on site
Minimise use of imported materials
Use materials from sustainably managed sources
Keep use of materials from non renewable sources to a minimum
Use low energy materials, keeping high embodied energy materials to a minimum
Use second hand/recycled materials where appropriate
Re-use existing buildings and structures instead of always assuming that new buildings are required

(b) Minimising External Pollution and Environmental Damage
for example
Design in harmonious relationship with the surroundings
Avoid destruction of natural habitats
Re-use rainwater on site
Treat and recycle waste water on site if possible
Try to minimise extraction of materials unless good environmental controls exist and avoid materials which produce damaging chemicals as a by product
Do not dump waste materials off site but re-use on site

(d) Minimising Internal Pollution and Damage to Health
for example
Use non toxic material, or low emission materials
Avoid fibres from insulation materials getting into the atmosphere
Ensure good natural ventilation
Reduce dust and allergens
Reduce impact of electromagnetic fields (EMFs)
Create positive character in the building and relationship with site
Involve users in design and management of building and evaluating environmental choices

1.2.1 Embodied Energy

An important principle in the above four principles is that of Embodied Energy. This is a topic of concern to many academics and researchers but as yet there is no internationally agreed method for calculating embodied energy. The term has already been mentioned in this chapter but it is worth examining it more closely as it is so central to the understanding of green building thinking. Essentially, calculating embodied energy enables one to evaluate the global rather than the local impact of particular materials and products. For instance an energy conscious householder may wish to install UPVC double glazing under the impression that this will be an environmentally friendly thing to do. However an embodied energy calculation might show that the energy used in manufacturing and transporting such windows was substantially more, over the life of the product than the energy saved in the house where it is installed over the same period. If one also takes into account the costs of disposal or recycling (if this is technically possible) and the environmental costs of disposing of toxic by products and so on, then other solutions to the windows, such as using timber might be more environmentally acceptable.

"calculations of embodied energy are complex, for they include the energy from the extraction of raw materials through to processing and erection. Taking transportation (as well as infrastructure) into account, not to mention a portion of the energy used to make mining, processing, transportation and construction equipment, one has a challenging task to arrive at a comprehensive single figure for the embodied energy of any given material. Considering the variety of materials which go into any building, a single figure for a building is even more daunting."[8]

Where do we draw the boundaries writes Thomas Keogh in a Masters Thesis at Queens University,

"should we consider the energy used to cook the building workers breakfasts?"[9]

At present there is no universally agreed basis for embodied energy calculations and experts either refuse to divulge their figures or disagree about exactly how many watts of energy are used to manufacture aluminium or transport hardwood from Malaysia. Until Government and European research agencies recognise the vital importance of supporting research in this field rather than simply funding new technology programmes, progress will be slow and information for the end user difficult to access. It will also be difficult to trust embodied energy figures produced by manufacturers unless there is an independent accreditation system to check them.

1.3 Why Green Building?

In order to understand the thinking behind green building principles it is necessary to remember why we should be so concerned with such issues in the construction industry. Perhaps producing **more** energy from renewable sources and protecting wildlife and habitats is much more important? Indeed there are many who do not give green building a high priority. It is surprising how many environmental groups, for instance, appear to attach a low priority to their built environment. Groups concerned with the natural environment, wildlife, habitats and so on, sometimes inhabit or build dreadful buildings using toxic materials and high embodied energy materials.

Many others see the issue purely in terms of energy efficiency or more specifically fuel efficiency and are largely unconcerned about the environmental impacts of the materials which they use to achieve reductions in gas, oil and electricity bills. Government and European research and development programmes such as Joule/Thermie, Save and Altener or the UK Clean Technology programme seem largely designed to encourage high technology development, leading to new and more products and systems which will expand industry and create new markets.

When the four main principles set out above are taken into account, it becomes clear that the building materials industry, the transport of materials and products, their construction on site and then the pollution and energy wastage coming from buildings collectively has a surprisingly wider impact on the environment than most other human activities. The Vales have suggested that 66% of total UK energy consumption is accounted for by buildings and building construction and services.[10] Thus the importance of buildings and the construction industry has to be seen as one of the most, if not the most important user of energy and resources in advanced society.

Major savings will not be achieved only by putting more insulation in homes or using low energy light bulbs, a much more fundamental review of all building materials production and construction methods, transportation etc. is required. Thus if we are concerned about ozone depletion, wastage of limited natural resources, such as oil, gas and minerals, the loss of forested areas, toxic chemical manufacture and emissions, destruction of natural habitats and so on, tackling the built environment is going to go a long way to addressing these issues.

1.4 How do you decide what is Green?

The question of how to decide what is or is not green is not easily answered. As has already been stated, there is no universal agreement on calculating embodied energy and numerous academics and professionals are devising environmental labelling and accreditation schemes in the hope that theirs will become the industry standard. The aim is to come up with a standardised set of criteria for environmental performance and provenance that will be internationally adopted and provide architects, manufacturers, builders and clients with a simple system for claiming that their building product or material is environmentally friendly. Many hope that this can be done using a simple numerical scale which incorporates all the issues such as embodied energy, emissions, toxicity and so on. Conferences have been convened to discuss this proposition and a number of systems have been devised to categorise or evaluate buildings.

Where does the Green Building Handbook stand in all of this? Some of our critics (not that there are many) say that the Green Building Handbook is flawed because our system of evaluating materials and products is not based on an independently agreed set of criteria. They say that it will take many years before the necessary scientific research and trans-national agreements have been reached before such agreed criteria can be established. Meanwhile the planet continues to be denuded of natural resources and pollution continues to pour into the water courses and atmosphere whilst these academic debates take place and demand for good advice on green building methods is growing. We thought it was better to get on and provide what information was available now instead of waiting for

the scientists and policy makers to agree, drawing on and digesting from authoritative published sources and the experience of practitioners in the field on our advisory group.

Also we have not attempted to create a standard system of classification so that users only need to apply a formula or simply give numbers to particular materials or products. Our aim has been to empower the user of the information to reach his or her own conclusion on the basis that they will do the best they can within the limits of current technology.

1.4.1 Environmental Classification Systems

Eventually, standards will be agreed that can be the basis of legislation, European and International standards. These will have the effect of forcing those less concerned with green issues to reduce their impact on the environment. Already building regulations have been improved to reduce energy consumption. Regulations also exist to reduce toxic emissions from building materials. But frequently such standards are watered down as a result of commercial pressures or fail to be properly enforced and will inevitably lag behind what is possible.

There are a number of environmental classification systems available or under development and more are likely to follow. Here is a list of some examples:

BREEAM (UK), the BRE Office Tool kit (UK), Home Energy Rating (UK), European Eco-labelling (Europe), Ecocerto(Italy), EcoLab(Netherlands), BREDEM(UK), SIB(Switzerland), BauBioDataBank (Germany), Waste/ Environmental Data Sheet (Europe), Athena (Canada), BEPAC (Canada), BMES Index (Australia) and probably many, many more.[11]

Plate 2. Exhibition of Green Building Materials at Construct 96, Belfast

Photo; Queens University Photographic Unit

These cover assessments of individual buildings, materials and products and a more detailed analysis of each of them might well be a suitable subject for a future issue of the digest. Some companies will make reference to having achieved British Standard (BS) 7750, which is a form of authentication that they have adopted some environmentally responsible practices and procedures, but these should not be taken as cast iron proof that they are not wasting energy or producing pollution.

Many larger building and development projects are now required by law to produce environmental impact statements (E.I.S.s) before planning permission is granted, but these documents, usually commissioned and paid for by the developers, cover broad issues of habitat impact and questions of planning law, but rarely, if ever, go into the detailed content or energy impact of the actual buildings themselves.

Eco-labelling seems to be the front runner for a system that will be commercially adopted and a UK Eco-labelling board, based in London, is now issuing guidelines to industry for this voluntary scheme for consumer products. Already a well known paint manufacturer has achieved an eco-label for one of their paints though the publicity for this does not make clear the basis on which it has been awarded.[12] The standards for paint eco-labelling are extremely complex and have been the subject of a great deal of debate and political horse trading at European level.

Reference to some of these systems will be found in the digest, where useful information could be gleaned or where claims were supported by references, but readers should beware of too much reliance on these labelling or assessment systems at present as most are in their infancy. In the digest we have tried to avoid relying on any one particular system of measurement or categorisation and those who wish to go more deeply into these issues will have to follow up the references.

1.4.2 Critiques of Environmental Assessment Systems

There is also a significant body of literature which is critical of current attempts to develop standardised systems of environmental criteria. Stephen Wozniak has argued that several assessment systems are flawed in that they rely on an uneven collection of criteria that are not based on any logical evaluation. Quite different methods of measurement are brought together into one system. Often really crucial environmental factors are left out simply because they couldn't fit them into the methodology.[13]

Elizabeth Shove has warned of the dangers of **standardisation** in that such an attitude in the past with public housing has, she says, led to a failure to take account of the

"cultural variability of building occupants and their creative, multi dimensional interaction with the built environment." [14]

In other words, such standardisation rules out opportunities for people to take responsibility for environmental standards and avoids variation between different circumstances. Rigidity can be dangerous as people fail to look behind the bland labelling to the criteria which have been used to formulate them. If something has an eco-label, it will be assumed to be o.k, when full awareness of the impact of the product may still lead many to question its use.

As the interaction of occupants with buildings, both in use and during construction can be a key factor in the environmental effectiveness of buildings, standards which leave out the human factor will inevitably be flawed. Attempts to produce standardised and systematised solutions to buildings have invariably led to problems because buildings are extremely complex, requiring creativity, imagination and judgement exercised in collaboration with clients and building users. For instance, the introduction of highly insulated, draught sealed buildings to save energy has led to severe problems of condensation and health problems for occupants as insufficient attention has been paid to ventilation. Attempts to compensate for this by introducing ventilation and heat recovery systems have ended up increasing the energy costs beyond the original reductions.[15]

Environmental classification systems which ignore the design and inter-relationship issues may not be successful. Unfortunately, many of the scientists involved in the eco-labelling and environmental criteria movement are in danger of overlooking this important lesson as the funding to develop such systems is inevitably going to come from the manufacturers of materials and the producers of building systems.

Not only are present environmental labelling systems scientifically underdeveloped, the ideology that underpins many of them pursues a purely neutral scientific goal, ignoring the social and political context in which buildings are produced. Indeed some guides ignore alternative materials and products, which are such an important feature of the handbook and green building practice.[16] In a purely commercial environment builders and developers may be concerned with creating the impression that they are being environmentally responsible whilst decisions about development and building procurement may be taken in a way that precludes proper consideration of environmental issues.

The danger is that many are concerned with finding a 'scientific', politically neutral, mathematical formula for awarding environmental credit points to particular materials, products and buildings, while making it possible for commercial manufacturers and developers to avoid the need to understand environmental issues themselves. The search for this holy grail could be futile in building, because an element of judgement and discretion will always be needed for each project which can take account of the inter-relationship between issues and the level of commitment to environmental action of the developer or client. It is unlikely that anyone will ever establish **absolute** standards.

1.4.3 Developing a flexible system of environmental guidelines

Rather than attempting to achieve a mathematical, politically neutral set of standards, which then hold up the danger of being applied in an inflexible way, what are required are guidelines based on scientific research against a whole range of questions that green designers and specifiers want answered. Judgement about what should and should not be used can then be made by well informed designers and clients through a process in which they take responsibility for the implications of their decisions. Unless designers and clients explore the issues, choices and implications of their decisions in a way that forces them to take responsibility for environmental impact, we will be avoiding the responsibility we all should exercise to use scarce resources wisely and protect the planet. Simply applying certain standards without investigating the reasoning behind them creates the danger of environmental criteria which are essentially cosmetic.

Of course this is a controversial point of view as there are many who believe that measures to protect the environment will never be taken unless stringent standards are applied through legislation. There is much to say in support of this point of view and indeed many of the issues referred to in the digest are a result of European legislation intended to protect industrial workers and the environment. Such base-line controls and requirements are necessary, but we cannot rely on legislation to determine behaviour. It is still necessary to change attitudes and this must be done through education of professionals and others in the construction industry in particular about the implications of specification decisions on the environment.

1.5 Limitations of Green Building

Another danger of eco-labelling and similar systems is that of 'green consumerism' where some people decide to adopt a 'green' life style which remains only superficial in its impact. Architects might similarly decide to adopt a green style of design without any fundamental concern for the underlying principles. Others may believe that by designing green buildings that this is, in itself, sufficient to solve the world's environmental problems.

Peter Dickens tells us of the danger of suggesting that

green design can "save the world", warning of a fetish of so-called environmentally friendly commodities which are simply new forms of consumer product. He argues for the need to change the relationships and processes which are causing the problem in the first place,
"to suggest that buildings and designs are themselves capable of creating sustainable societies could be, to say the least, seriously misleading."[17]

Thus simply having a green image is dangerous without any understanding of where materials and products come from, how they are manufactured and the impact that this has on society. It is how materials and products are produced and then used, rather than simply being labelled green that makes the difference in green building practice.

Photo-voltaic cells, for instance can generate electricity using the power of the sun and reducing our dependence on fossil fuels, but they are currently very expensive to produce and can hugely increase the embodied energy costs of a building. Such costs are likely to reduce dramatically over the next few years, but at present other solutions may be a lot less hi-tech and flashy, but just as effective in reducing heating and electricity costs.[18] Importing green products from around the world can also be hard to justify, unless such a practice is making a sustainable contribution to a particular economy and can be justified in global terms.

1.6 The Handbook methodology.

Given the problems outlined above, the green building handbook does not try to offer a simplistic and easy set of conclusions because any serious green designer would not use it in this way. Instead it digests material that has already been published or gleaned from the panel of advisors and presents it in a way that leaves the reader with the job of coming to his or her own conclusions. It is not claimed to be definitive, providing a brief overview of the information currently available.

A pull out sheet summarises the terms used in the product table and this can be referred to until the reader becomes familiar with the tables. The product tables are used with a scale of zero to four set against a number of headings. No attempt has been made to attach a particular weight to each heading and the size of blob is simply a handy way to help the eye scan the table. Reasons are then given for the size of each blob and the reader is free to accept or question these. As more research is done and more published on the subject it will become easier to establish agreed figures for issues such as embodied energy, including the energy used in manufacturing, packaging and transport etc., but even this can vary from one project to another. Headings used include: a unit price multiplier (based on an estimated life cycle costs over 60 years), Production

Plate 3. Calthorpe Community Centre Timber Frame Building, Kings Cross London

Photo: Architype

Impacts, Embodied Energy, Resources (Bio), Resources (Non Bio), Global Warming, Toxics, Acid Rain and Photo chemical Smog, Post production Impacts, Thermal Performance, Health Hazards and Recyclability.

Each digest or chapter attempts to give some guidance under each of the above headings, when these are relevant, without any pretence that a magic number can be allocated under each. In many cases the categorisation is based on the judgement of the researchers from the best available information. Where such judgement can be questioned indicates not the failure of the Green Building Handbook so much as the need for further research to be undertaken on the issue in question.

Much of the literature on the subject tends to confuse means and ends and sometimes promotes a particular approach such as timber framing as being the only way to achieve green buildings. The Handbook does not attempt to cover construction systems at this stage though this would be appropriate for future issues. The holistic and complex nature of green building design would involve a great deal of illustrations to fully explain, however a few examples are illustrated in the following chapter to give some idea of what can be produced when green building principles are followed.

Each chapter takes a small piece of the jig-saw by concentrating on a particular product, material or building element and provides as much information as possible that may enable the designer or specifier to make realistic decisions on how to work towards the four main objectives listed on page 6.

It is important to note that the Handbook is essentially pragmatic in practice. Almost all built development, indeed much human activity consumes resources and affects the environment. Those concerned with green building wish to carry out their activities in a way which does as much as possible to protect the environment and the health and welfare of building users. At present we have to do this in a society which is largely unconcerned with such issues and a construction industry which, even at its most advanced as in Germany or the Pacific North West of the USA regards green building as only a 20% share of the market. Sadly we don't start with a clean sheet, we start with a very dirty sheet and in the short term the objective is to clean up our act. Going any further raises more fundamental issues about life and consumption.

The construction of a few green buildings or the limited adoption of some environmental controls should not encourage us to become complacent, instead such steps forward should be signposts to addressing more fundamental issues in the future.

The Green Building Handbook has only modest ambitions to provide information, not only to professionals and experts but to a wider audience so that a better informed society is making demands for buildings which do not damage their health or the environment. The Handbook is firmly rooted in the ideas of the community technical aid movement, ie, that ordinary people in community and voluntary groups should be able to take decisions about their environment and that information and participatory processes should be used to ensure that they are able to make decisions that are in their best interests. While there are a growing number of prestige buildings with claims to be environmentally friendly in their approach, many thousands of people are homeless or living in damp, badly insulated homes, or work in unhealthy, poorly ventilated, poorly lit environments. The real test of the usefulness of the Handbook will be if the information is put to use enabling and empowering working class and unemployed and disadvantaged groups to get a better environment. Thus the intention of the Handbook is not to reinforce the power of experts and professionals but to demystify technical knowledge so that it is accessible through the technical aid movement and professionals who are committed to user participation. Also by making it clear that there are no easy answers or simple eco-labelling systems, there must be debate and discussion on each project so that those involved take responsibility for the environmental impact of their decisions.

The handbook therefore does not lay down strict do's or don'ts but instead respects the users of the digest who will have to investigate the complexities of each project, using the handbook where it is relevant.

1.7 References

1. The Blind Watchmaker (Richard Dawkins) Penguin Books, Middlesex, 1988
2. " A House in the Country" *(Deyan Sudjic) The Guardian June 21, 1996
3. Green Architecture (Robert and Brenda Vale) Thames and Hudson London, 1991
4. Greener Buildings, The Environmental Impact of Property (Stuart Johnson) The Macmillan Press London, 1993
5. Ecological Design (Sim Van der Ryn & Stuart Cowan) Island Press Washington DC, 1996
6. B.S.R.I.A. 1996 Circular from Steve Kilford
7. Sustainable Development Services, 1996 Practice Profile Seattle
8. Recycling building materials in Ireland (Thomas Keogh) Queens University Belfast Department of Architecture Unpublished MSc Thesis, 1996
9. Op Cit.
10. Towards a Green Architecture (Robert and Brenda Vale) RIBA Publications London, 1994
11. Eco-labelling in Europe Conference Report (Howard Liddell (Ed.)) Robert Gordon University Aberdeen 1995
12. UK Ecolabelling Board Newsletter No.10, London
13. Environmental Assessment of Buildings and Building Developments - A Logical Methodology for the World. (Stephen Wozniak) Mimeo Bedmond Herts, 1993
14. Set Points and Sticking Points for Sustainable Building Design (Elizabeth Shove) in Cole R.J.. (Ed.) Linking and Prioritising Environmental Criteria Toronto School of Architecture, University of British Columbia, 1995
15. European Commission Joint Research Centre- Environment Institute 1996 Indoor Air Quality and the Use of Energy in Buildings Report No. 17 (EUR 16367 EN) Luxembourg
16. The Green Guide to Specification - An Environmental Profiling System for Building Materials and Components. (David Shiers and Nigel Howard) Post Office Property Holdings. 1996
17. Architecture as Commodity Fetishism, Some Cautionary Comments on Green Design. (Peter Dickens) Housing Studies Vol.8 No.2 pp 148-152. 1995
18. Unlocking the Power of Our Cities - Solar Power and Commercial Buildings Greenpeace, Canonbury Villas London, 1995

How to Set About Green Building   2

2.1 Using the Handbook

The Green Building Handbook is not a design manual nor does it lay down rigidly fixed standards or guidelines. It is intended to sit next to the drawing board or in the technical library and to be regularly consulted by designers specifiers and clients. It will need to be supplemented and up-dated, particularly as knowledge of the environmental impact of materials is growing at a rapid rate and as more 'alternative' products become available.

It can also be used as part of a discussion between clients and their consultants when the difficult decisions about choices of materials and finishes are made. Without an awareness of environmental issues, decisions are usually cost led, with the cheapest being preferred in most cases. The Green approach, however has to take into account the lifetime and environmental costs of decisions and this means that the lowest initial cost is not always the best. This does not mean that Green buildings need to be more expensive. Often the Green choice can actually reduce initial costs, but even where they are increased, savings can be made over the life time of the building.

Many Green specification choices are more expensive than conventional alternatives at present, because the market is so small for Green products, but as consumer demand changes prices will come down. Often Green products are inherently cheaper as they used recycled or by-product materials, however they are nor always widely available and the industry is still trying to off-load toxic and environmentally damaging materials.

2.1.1 Costing the Earth?

Weighing up choices about specification involves assessing the payback in terms of savings of energy and reducing environmental damage. We all have a responsibility to make these calculations, but we also need to consider how long materials will last and what will happen to them when they are taken down or thrown away. Life cycle costing is a well established concept, though rarely applied when short term considerations are so universally prevalent. However if the environmental impact and embodied energy costs are brought into the analysis , the picture changes radically. As taxes and other government controls make the producer and consumer pay the real environmental costs, attitudes are still slow to change. The introduction of the landfill tax has created many new environmental initiatives but has not yet significantly altered waste practices. However a new recycling industry is slowly developing.

The environmental costs of dumping waste are very high, and as these become more immediate the construction industry is already changing its practices. Building sites no longer need to be covered in rubbish and waste materials can be sorted tidily and re-used. Concrete and steel from

demolished buildings is now carefully extracted and recycled and the technology to make this possible is readily available and cost effective. Specification decisions need to take into account what will happen when the building needs maintenance, or elements have to be replaced . Can the materials be recycled , will this reduce long term costs , and so on ? Costing of embodied energy can highlight the advantages of re-use, rehabilitation and conversion of existing buildings, normally considered to be more expensive than demolition and new build.

2.1.2 Strategic Approach to Design

Examining the environmental impact of design and specification decisions can affect the whole strategy when considering building options. Whether to build new or

Plate 5. Straw Bale House

Photo: Queens Uiversity of Belfast Photographic Unit

convert an old building or whether the activities need a building at all . Where should it be sited, in an existing settlement or out in the country generating the need for more traffic, or could public transport provision influence decisions? At present only larger projects tend to require an environmental impact statement, but eventually, all building proposals should, perhaps, be required to assess their environmental impact, beyond the present, relatively low energy saving requirements in the building regulations. This would ensure that green issues would be taken into account in all projects.

Unfortunately it is often assumed that a building can be designed and then someone can come along to make it energy efficient or green at a later stage. This is a mistaken approach as initial strategic and design decisions may rule out green principles. Anyone with a building or considering development should seek advice about environmental issues right from the beginning.

2.1.3 Importance of Design

While the Green Building Digest can be a useful reference to inform decisions it is necessary to employ consultants who fully understand how to integrate all design decisions because it is the interrelationship between all the elements of a building that can determine its success in environmental terms. It is also important to understand that assembling a collection of green materials will not necessarily result in a successful building. The importance of good design cannot be under-estimated. It is essential to establish a good partnership where all are committed to the idea of a green building and are willing to take responsibility for the decisions that follow. With a big building requiring architects, structural engineers, quantity surveyors and mechanical and electrical engineers, it is essential that the whole team are involved in the design process from the beginning and are equally committed to the green approach. It is pointless designing a building where the M&E consultants, for instance, come along later and put in a mechanical air conditioning system once the building is largely designed and yet this frequently happens. Alternatives such as natural ventilation need to be considered from the beginning.

Even with a small building, such as a house, environmental issues must be considered from the beginning of the process. It is not unusual to be contacted by clients who have appointed an architect to design an environmentally friendly building, who then ask for green experts to come along and give the consultant a crash course in how to turn their design into something that might be a little less damaging. Green building is not a separate specialist discipline, it must be fully integrated with the whole process.

2.1.4 Getting the right advice.

Clients need to appoint consultants who fully understand how to undertake the green building task and have some experience of working in this way. They should also visit examples of good practice and this may well involve some travelling. The few good examples of green building practice are fairly thinly spread throughout the UK, Europe and the USA. In most cases not every problem has been solved or even attempted, so you may have to visit several projects to see everything you are interested in.

Finding architects and other consultants who understand green principles is getting easier and clients are best advised to contact one or more of the organisations listed in this book for lists of professionals in their area rather than using the establishment professional bodies.

Finding builders who are familiar with green building methods is also not too easy. The Association of Environment Conscious Builders may be able to put you in touch with someone, but you may have to educate more conventional companies. Even where lip service is paid to environmental principles, the construction industry is notoriously conservative and workers on site will need constant supervision and even re-training. They will not be able to understand why you don't want them to coat all the timber in toxic preservatives and they are used to wasting as much as 10% of material on site. High levels of insulation can be pointless if careless building leaves cold bridges and gaps. These will not be apparent once everything is covered up.

2.1.5 Sourcing Green Materials

Obtaining green materials will also be difficult even with the details of alternative suppliers given in the digest and other publications. You may not want to source material from far afield as transport costs will put up the embodied energy costs and many of the best ecological products come from Scandinavia and Germany. Often more conventional builders merchants will not welcome enquiries about the source of materials such as timber and will not be able to advise you on the nature of toxic emissions from different products. Others can appear to be environmentally aware and will tell you that their timber or products are from sustainable or renewable sources. However a simple question as to what accreditation or authentication system has been used will frequently produce a blank stare !

Anyone setting out with the intention of creating a green building will be helping to blaze a trail and the greater demand for environmentally friendly materials and products, the easier they will be to obtain. Market pressures as well as changing public sector policies about specification will make green building easier. There are signs that the industry is rapidly becoming aware of the demand for environmentally friendly products and while only one paint product, so far has a Eco-Label, many more are likely to follow.

Examples of Green Building 3

3.1 Examples of Green Buildings

One of the best ways to understand and see the benefits of green building is to visit or read about examples. Each will illustrate a range of attempts to achieve green results. In most cases, good designers will make it clear that they have not been able to achieve everything that they intended and that the building may not be as green as they like. This is not surprising when green building is such a new idea. Each new project allows the boat to be pushed out a little further and experimental techniques to be tried and tested. Gradually the accumulation of knowledge and experience will make it easier to improve good practice in the future.

We have selected a handful of buildings which show a range of green building ideas in practice. Only a brief account is given of each project, though future editions of the Digest may include more detailed case studies and working drawings. There are many other examples of green buildings, not illustrated here both in the UK and further afield and the potential green builder will need to learn from a wide range of examples to see all possible approaches and techniques tried out.

3.1.1 Chapel Allerton Leeds

This is a self build project of three terraced houses in a suburban area of Leeds. The scheme is intended to put into practice a wide range of green principles and is one of the best examples of an holistic approach to be found in the UK. The self builders are now in residence and will not necessarily welcome constant visits from curious onlookers so they are going to run occasional courses to demonstrate the various features of the houses.

(a) Principal Features

Modified "Walter Segal" timber frame system of structure. Timber was treated with borax to avoid normal toxic chemicals. All other toxic materials have been eliminated as far as possible. A considerable amount of second hand materials have been used as well as locally sourced green timber.

High levels of recycled paper insulation have been used (150mm in the walls and 300mm in the roof) and there are some passive solar gains. The street front roof uses pan tiles and was designed to appease the planning authority but at the back a planted roof and timber cladding present a greener solution. Each house has a composting toilet, designed by the builders rather than buying an expensive off the shelf version and 'grey' water from sinks and showers will be recycled through reed beds in the back garden. All rainwater will be collected off the roofs with the aid of timber gutters and then purified and used in the building again using system designed for this scheme making them almost entirely independent of mains water. The scheme has a mortgage from the Ecology Building Society.

For further information contact Jonathon Lindh or Matthew Hill Leeds Environmental Design Associates
1 Grosvenor Terrace Leeds LS6 2DY
Telephone 0113 278 5341

Plate 6. Housing at Chapel Allerton, Leeds (Front Elevation)
Photo: H. Salt

Plate 7. Housing at Chapel Allerton (Rear elevation)

Photo: H. Salt

Plate 8. Centre for Understanding the Environment,   Horniman Museum,
Dulwich, South London.

Photo: Architype Ltd

3.1.2 Centre for Understanding the Environment,

Horniman Museum,
Dulwich, South London.

This extension to the Horniman Museum was built to provide educational facilities and displays on local and global environmental issues It was decided that the building itself should be a model of green thinking, displaying energy efficient technology and using environmentally friendly materials and construction techniques. The building, therefore, is a green exhibit in itself.

Constructed largely of timber, obtained from sustainable sources, the structure includes highly innovative triangular timber beams which support the floor and roofs and also serve as ventilation ducts. Recycled newsprint insulation and breathing wall constriction have been used. There is a green roof and water is recycled through a system of ponds which are part of the habitat around the building. A passive ventilation system is used and air is pre-heated in the floor in winter, though heating bills are claimed to be as low as £250 per year. All paints and other treatments were using non toxic , organic paints.

Contact : The Horniman Museum Telephone 0181 699 1872
or Architype Design Co-operative
4-6 The Hop Exchange, 24 Southwark Street   London SE1 1TY
Telephone   0171 403 2889

Plate 9. Centre for Understanding the Environment (Internal),   Horniman Museum, Dulwich, South London.

Photo: Architype Ltd

Plate 10. House in Sligo, Republic of Ireland

Photo: W. Rothwell

3.1.3 House in Sligo, Republic of Ireland

This is a self built house constructed over 2-3 years using a modified Walter Segal Timber Frame construction system. There are three main frames which were put up over a weekend with the posts built off concrete pads to try and cope with the West coast of Ireland driving rain. One of the advantages of the Segal post and beam frame system is that major alterations can easily be made and in this case the area of window on the south facade was changed without any additional cost after the main walls were up!

Recycled newsprint insulation has been used in the walls and roof . the walls are lined with plywood to give extra rigidity, compromising the breathing wall concept somewhat and the architect might use a different system today . Much of the timber was sourced from forests which claimed to employ sustainable planting strategies, in Ireland and treated with Borax. Even the cedar cladding came from Ireland.

Roofing slates were Irish grey green slates recovered from a building which was being demolished in Sligo town. Internal walls are finished with lime. A heat recovery ventilation system has been installed and annual heating costs are proving to be extremely low.

Contact Colin Bell Architect
13 Johnson's Court, Sligo, Republic of Ireland
Telephone 071 46899

Plate 11. House in County Laois, Republic of Ireland.

Photo: W. Rothwell

3.1.4 "Taliesin" House in County Laois Republic of Ireland

This is a conversion of an old stone barn in countryside to the South East of Dublin. Because of the environmental requirements to achieve high levels of insulation , the house is constructed from a timber frame, inside the stone shell. The frame is entirely independent of the old walls , with a cavity between the new construction and the old. Any materials such as the slates were carefully recycled . Timber cladding above the stone walls 'hangs' off the roof structure .

This made it possible to use breathing wall and roof technology with recycled paper insulation . Timber was obtained from Irish sustainable sources . Water based paints and stains were used throughout. The final result has a timeless quality with even the modern extension and conservatory sitting in harmony with the existing building.
Contact Michael Rice Architect
Rossleaghan County Laois
Republic of Ireland
Telephone 0502 22747

3.1.5 The Old Mill Crossgar, N.Ireland

This is an example of where an existing redundant structure, in this case a disused and derelict mill and a barn, have been converted to provide a house. There are many such redundant agricultural and industrial buildings in towns and the countryside and it is a waste of energy and materials for these to be demolished and cleared to landfill sites unless the buildings are in dangerous condition.. Often there are planning restrictions on the use of such buildings which can mean that they are allowed to sit empty and decay, thus robbing the environment of reminders of the past. Often such buildings are made of stone or brick using techniques which would be uneconomical to replicate today . There can be some opposition to conversion from environmentalists who are trying to protect wildlife habitats for bats or Barn Owls, others would like to see buildings such as mills preserved as working mills rather than converted into houses. Such factors should be carefully thought through before conversion is pursued. Sadly many such buildings are now demolished to feed the fashionable demand for second hand materials so these contradictory issues have to weighed carefully.

The Crossgar project is an experiment in how to re-use such structures and involves several phases. There are environmental problems in retaining existing structures as it may be hard to achieve the high levels of insulation expected in green buildings, but even where this is not achieved, overall embodied energy savings may justify such projects. In the Crossgar example, a high level of second hand materials have been used including polystyrene insulation rescued from a cold store which was being demolished. Second hand and locally sourced timber has been used . Heating is from a gas boiler and some passive solar benefit is derived from a conservatory. A passive ventilation system , rainwater recycling and grey water reed bed treatment are part of Phase 2 . Phase 1 included a grass roofed garage/workshop building constructed entirely by inexperienced architecture students. A key policy on this project was to avoid wastage and every scrap of wood and stone has been reused on site in some way. Not a single skip was used to take waste material off site.

Contact Rachel Bevan Architects 45 St. Patricks Avenue Downpatrick Northern Ireland Telephone 01396 616881

Plate 12. Old Mill, Crossgar (South & East elevations)

Photo: T. Woolley

3.1.6 Linacre Collage Oxford

This is a student residential building in Oxford, part of a post graduate college on the edge of the Cherwell Meadows. The College decided to make the building a showcase of green ideas and has been successful in that the building won the Green Building of the Year Award 1996. While the somewhat reproduction style of architecture meant that the building has used brick and a substantial amount of concrete in the basement, the college took the remarkable step of buying an area of Tasmanian Eucalyptus Forest to offset any criticism of CO_2 emissions which they had not been able to prevent. Other tree planting to replace trees that had to be cut down, was carried out in Oxford and the design incorporates nesting boxes and other nature conservation features. The embodied energy costs of the building were carefully worked out and this influenced the choice and specification of materials. Recycled newsprint insulation was used with mineral fibre in the basement. Timber windows were selected and plastics materials excluded. Demolition materials were used for hard-core. Low formaldehyde materials and non toxic paints were also selected.

A highly energy efficient heating and heat recovery system was used with passive stack ventilation. There have been problems with a waste water recycling system but these are being remedied. Careful auditing and calculations and monitoring mean that the success of this building in environmental terms can be evaluated if the results are published.

Part of the attraction of the college as an example of green building is that while including a number of state of the art 'eco' features, the architects have maintained a traditional aesthetic, in contrast to the high tech or rustic extremes of many green buildings

Contact ECD Architects
11 Emerald Street London WC1N 3QL Telephone 0171 405 3121

Plate 13. Linacre College, Oxford

Photo: W. Rothwell

Part 2

Product Analysis & Materials Specification

How to Use the Handbook

Life Cycle Analysis

The Green Building Handbook's Product Tables present a summary of the environmental impact of each product covered in an 'easy-to-read' format. A circle in a column will indicate that we have discovered published comment on a particular aspect of a product's impact. The larger the circle the worse an environmental impact is thought to be (in the opinion of the author). Marks on each Table will only indicate poor records relative to other products on the same Table.

Every mark on the Product Table has a corresponding entry in the Product Analysis section, which explains why each mark was made against each particular product.
Life Cycle or 'cradle-to-grave' analysis of a product's environmental impact is a relatively new, and still contentious field. It is accepted that it should involve all parts of a product's life; extraction, production, distribution, use and disposal. The Green Building Handbook's Product Tables amalgamate these for ease of presentation, so that issues involving the first three, extraction, production and distribution are presented in the nine columns grouped under the heading 'Production'; the last two, use and disposal, are presented together under the heading 'Use'.

Less well accepted are the more detailed headings under which life cycle analysis is performed. Those we have used are based on those used by other LCA professionals, but developed specifically for this particular use - presenting information about building products in a simple table format.

The most fundamental problem with LCA is in trying to come up with a single aggregate 'score' for each product. This would entail trying to judge the relative importance of, for example, 50g emission of ozone depleting CFC with a hard-to-quantify destruction of wildlife habitat. In the end the balancing of these different factors is a political rather than scientific matter.

Key to Product Table Ratings

The environmental impacts of products are rated on a scale from zero to 4 under each impact category. A blank represents a zero score, meaning we have found no evidence of significant impact in this category. Where a score is assigned, bear in mind that the scores are judged relative to the other products on the same Table.
The following symbols represent the impact scale:

⬤ ... **worst or biggest impact**

● .... **next biggest impact**

• ..... **lesser impact**

· ...... **smaller but significant impact**

[blank] **no significant impact**

Key to Product Table Headings

Key to Product Table Ratings

The environmental impacts of products are rated on a scale from zero to 4 under each impact category. A blank represents a zero score, meaning we have found no evidence of significant impact in this category. Where a score is assigned, bear in mind that the scores are judged relative to the other products on the same Table.

The following symbols represent the impact scale:

⬤ ... **worst or biggest impact**

● .... **next biggest impact**

• ..... **lesser impact**

· ...... **smaller but significant impact**

[blank] **no significant impact**

Unit Price Multiplier
This column shows the relative cost of the different options listed on the table based on a standard unit measure.

Production
This group heading covers the extraction, processing, production and distribution of a product.

Energy Use
More than 5% of the UK's total energy expenditure goes on the production and distribution of building materials. This energy is almost always in the form of non-renewable fossil fuels.
In the absence of information on other aspects of a product's environmental impact , energy use is often taken to be an indicator of the total environmental impact.

Resource Depletion (biological)
Biological resources, whether of timber in tropical forests or of productive land at home, can all be destroyed by industrial activity. These can only be counted as renewable resources if they are actually being renewed at the same rate as their depletion.

Resource Depletion (non-biological)
Non-biological resources are necessarily non-renewable, and so are in limited supply for future generations, if not already. These include all minerals dug from the ground or the sea bed.

Global Warming
Global warming by the greenhouse effect is caused chiefly by the emission of carbon dioxide, CFCs, nitrous oxides and methane.

Ozone Depletion
The use of CFCs and other ozone-depleting gases in industrial processes still continues despite many practicable alternatives.

Toxics
Toxic emissions, to land, water or air, can have serious environmental effects, none of which can ever be completely traced or understood.

Acid Rain
A serious environmental problem, causing damage to ecosystems and to the built environment. Caused mainly by emissions of the oxides of sulphur and nitrogen.

Photochemical Oxidants
The cause of modern-day smog, and low-level ozone, causing damage to vegetation, materials and human health. Hydrocarbon and nitrogen oxide emissions are chiefly responsible.

Other
No 'check-list' can ever cover all aspects of environmental impact. See the specific Product Analysis section for an explanation of each case under this heading.

Use
This group heading covers the application at the site, the subsequent in-situ life and the final disposal of a product.

Energy Use
Nearly 50% of the UK's total energy consumption is in heating, lighting and otherwise serving buildings. The potential impact, and therefore potential savings, are enormous.

Durability/Maintenance
A product that is short lived or needs frequent maintenance causes more impact than one built to last.

Recycling/Reuse/Disposal
When a building finally has to be altered or demolished, the overall environmental impact of a product is significantly affected by whether or not it can and will be re-used, repaired or recycled, or if it will bio-degrade.

Health Hazards
Certain products cause concerns about their health effects either during building, in use or after.

Other
Again no list like this can ever be complete. See the specific Product Analysis section for an explanation of each case.

Alert
Anything that we feel deserves special emphasis, or that we have come across in the literature that is not dealt with elsewhere, is listed here on the Table.

Energy 4

4.1 Scope of this Chapter

This chapter looks at the environmental impact of the choice of energy supply for space and water heating in buildings. Both conventional fossil-fuel energy and alternative renewable sources are considered, together with the practicalities of alternative renewable sources. Different types of equipment (e.g. boilers, generators, etc.) are not covered in any depth.

4.2 Introduction

"The last two decades have witnessed scientific consensus that the burning of fossil fuels has to be capped and eventually reduced."

(M K Tolba (3))

Global warming, the rise in global average air temperatures caused by increased emission of greenhouse gases, is likely to cause such changes in climate, weather patterns and sea levels that the lives of everyone on earth will be affected. Its primary cause, though there are others, is the CO_2 emissions from the burning of the fossil fuels - coal, oil and gas.

These fossil fuels are themselves in limited supply. Optimistic estimates for world crude oil production expect that it will peak within the next five years, and from thereafter be in decline.[1] We cannot rely, however, on the oil wells (or gas wells or coal mines) drying up in time to stop global warming. Action to reduce the use of fossil fuels must begin now.

Why Me?

As a building designer, specifier or in any other capacity responsible for the choice of heating systems in a building, you have more opportunity to do something about global warming than anyone except the politicians and energy multinationals. The use of energy in buildings in the UK is responsible for just over half of the country's total CO_2 emissions, twice as much as that from industry or transport.[1,12] 40% of UK energy is expended on heating buildings,[7] 25% on heating our homes.[12]

If any impact is to be made on global warming, drastic changes will have to be made in the way we heat our buildings. (See below, 'A safe level for CO_2 emissions?') To put it in perspective, the embodied energy of building materials, a key issue in other chapters, is perhaps ten or a hundred times less important for global warming than burning fuels for heating buildings.[14]

A safe level for CO_2 emissions?

One estimate reckons that cuts in CO_2 emissions of around 75% must be made in the industrialised countries if the effects of global warming are to be arrested.[14]

According to Friends of the Earth, "no definitive answer can be given to the question of what levels of reductions in emissions [of greenhouse gases] will be required to keep future climate change within tolerable limits... It is clear... that a short term commitment of all industrialised nations to carbon dioxide reductions in the order of 25-50 per cent by year 2005 is needed if the risks of climate change are to be minimised."[21] That is just in the next eight years.

It must be emphasised that the science behind these figures is still subject to much debate, but in view of the probable threat, it would be prudent to adopt the precautionary principle and act to substantially reduce CO_2 emissions.

The governments of the most progressive countries on this issue (e.g. Austria, Germany and New Zealand), are currently committed to around 20-25%. In the longer term, cuts would have to go further - 45-55% by 2050, possibly 80-100% by 2100. These figures also assume reductions of other greenhouse gases, and an end to deforestation.

Friends of the Earth's medium term view is that "an appropriate initial CO_2 reduction target for an industrialised country such as the UK would be at least 30 per cent by 2005 on 1990 levels."[20] Given the slow rate of change of the nation's building stock, an achievement of that sort of reduction in the energy used by buildings by that date, would necessarily involve improvements to existing buildings.

Even if all new buildings were designed and built for zero CO_2 emissions, this would not, on its own, be enough to meet this target.

The Truth About Being Economical

The supply of energy can either be capital intensive, as in the case of wind power, where a high initial investment is needed, but the running costs are minimal, or expenditure intensive, as is the case for most fossil fuels where the running costs (fuel) dominate. It would seem to be common sense that for any long-term project (such as heating a building) capital intensive, low-running-cost options should always work out cheaper in the long run, and therefore be generally the most widely used.

That this is not the case is due, in part, to the short term-ism of the suppliers of capital. It is also due to the effects of judging these costs against the benchmark of the marginal cost of using the equivalent amount of the cheapest fuel, usually gas or oil, and the fact that the true costs of these options are seriously underrated if their full environmental costs were also to be included.[36]

What People Want

Energy is not, of course, what people want. They want the services it can provide: comfort and warmth; hot water for washing; light to see by.[7] This is shown in the statistics for comfort standards in dwellings. From 1970 to 1991 average temperatures in dwellings rose from 12.83° to 16.66°C, but there was no great increase in consumption of energy for space heating. It came about due to the increased use of more efficient appliances - dwellings with central heating rose from 34% to over 80%.[11]

A BRE study has shown that proven technologies could reduce CO_2 emissions from the energy use of existing dwellings by 35% without reducing comfort levels. Of these improvements, two thirds were achievable by increased insulation, one third by increased appliance efficiency. (The effect of switching fuels was not included.) Over two thirds would actually save money.[12]

Any Answers

The answer to global warming and resource depletion is simple. Firstly energy requirements should be minimised by good design. Then, wherever possible we should use solar-based renewables, such as sun, wind or biomass, rather than fossil fuels, and where we can't, the efficiency with which we use fossil fuels should be maximised.[35]

This 'wherever possible' is important. A key strategy for sustainable development is the appropriate use of resources and technology. An ideal policy[†] would, until realistic replacements arrive, limit the use of fossil fuels thus: oil (as petrol, diesel etc.) should be reserved for mobile transport fuel; gas for high temperature industrial

---

The Political Climate

Under the United Nations Framework Convention on Climate Change, the UK is committed to stabilising greenhouse gas emissions "at a level that would prevent dangerous anthropogenic interference with the climate system". Also, as a developed country, the convention commits us to returning emissions of each greenhouse gas to 1990 levels by 2000. European Environment and Energy ministers also agreed that the community as a whole should stabilise CO_2 emissions by 2000 at 1990 levels.

The UN commitment is the strongest, and is aimed at sustainability. Even if industrialised nations stabilise at 1990 levels, global warming could still rise at twice the rate considered tolerable by sensitive ecosystems and vulnerable populations.

---

processes; coal for electricity generation for lighting, communication, stationery machinery and tracked transport.[19] Nowhere does the heating of buildings come into this list, apart from maybe some surplus heat from the power station via a district heating system. Crucially, realistic replacements are here already for heating buildings, but not so for the other uses of fossil fuels. It therefore falls on the building sector to take on more than its 'fair share', and to take the lead in renouncing fossil fuels.

---

† N.B. This strategy only deals with global warming and resource depletion issues - localised pollution effects from fossil fuel burning are a further complicating factor.

"On the scale of human history, the era of fossil fuels will be a short blip." (Meadows et al. (35))	Fuel Cost (p/useful kWh)	Resource Life	Global Warming	Acid Rain	Photochemical Smog	Particulates	Toxics	Risks	Other	Comments
Fossil Fuels										
Coal	2.5 - 3.2	•	●	●	●	●	●	●	●	Modern large-scale coal burning can be much cleaner than small scale boilers and fires.
Oil	2.0	●	●	•	•	●	●	●	●	The Brent Spar episode alerts us to the toxic wastes potentially associated with oil extraction.
Gas (North Sea)	1.7 - 2.3	●	•	•	•		•	●	●	Natural gas (methane) is itself a greenhouse gas.
LPG	2.6 - 4.1	●	•	•	•	•	●	●	●	Almost as clean as natural gas, but transport is usually by road rather than pipe.
Electricity (Grid, with current supply mix)	4.2 - 8.6	•	●	●	•	●	●	●	●	The 'energy mix' in the grid of coal, nuclear, oil, hydro, wind etc will, hopefully, get cleaner over time as more renewables are used.
Renewables										
Solar, Wind etc	0									alert: some evacuated tube collectors may contain ozone depleting gases. Also PV toxics such as cadmium.
Wood and other biomass	0 - 2.1		•	•	●	●	●	•	•	This is the 'worst case scenario' for biomass - modern stoves/boilers may be cleaner.

4.3 Best Buys

or What You Can Do...

- Remember the cheapest form of energy is conservation - minimise energy requirement by good design.
- Passive solar design is by far the best environmental option for space heating.
- Don't leave it to the end to 'bolt on' a heating system, include it from the beginning of the design stage (especially so with passive solar).
- Active solar for water heating is now a proven technology, and though expensive, can give worthwhile contributions to water heating.
- Maximise use of renewable sources wherever possible - e.g. a wood stove for backup heat in the winter.
- If you must use a backup fossil fuel, use gas (or LPG if not on mains) and a condensing boiler, with good controls and meters.
- District or community heating systems offer potential for economies of scale that make renewables such as active solar for space heating, or even wind and biogas, worthwhile.

"We will inevitably achieve 100% reliance on renewable sources eventually. All that we can determine is how quickly we move towards that goal."[13]

4.4 Product Analysis
4.4.1 Fossil Fuels

(a) Coal

Resource Life

Estimates for the life of the world's coal reserves vary between around 200 to 500 years.[1,35] It is thus by far the most abundant fossil fuel. Nevertheless, it is still a finite resource, and as the other fossil fuels run out, is likely to be used in increasing amounts, so its lifetime could be much shorter.

Global Warming

Burning coal releases more CO_2 than other fossil fuels,[6,11] and the deep mining of coal releases methane, another potent greenhouse gas, to the atmosphere.[3,22]

Acid Rain

Coal naturally contains sulphur in varying amounts depending on the grade of the coal and where it comes from. It is this sulphur content that gives rise to SOx the chief acid rain-forming gases. Coal burning is the major cause of acid rain, causing around 75% of SO_2 emissions in the UK.[22]

Photochemical Smog

Coal burning is responsible for significant quantities of oxides of nitrogen, hydrocarbons and carbon monoxide, the photochemical smog gases.[3,22,34]

Particulates

The particulate emissions from coal burning were well known in British cities before the Clean Air Acts cleaned up the famous smogs. Particulates from coal burning are around 3 times the quantity as from oil.[3,34]

Toxics

Coal smoke contains a wide range of harmful chemicals, some of which are carcinogenic.[16,33] Trace amounts of radionuclides (radioactive elements) are present in coal, and are released on combustion.[3]

Risks

The occupational risks of deep coal mining, both to the miners' health and from accidents and cave-ins are well known.[33] The use of open fires in the home also increases the risk of house fires.[33]

Other

The impact of coal mining on the local environment can be considerable: impacts include land use for mines and spoil heaps; subsidence; disturbance of habitats by open cast (surface) mining; pollution of water courses and tables by acid and salted mine drainage, water wash treatment and runoff from storage heaps; transport of coal by road; dust emissions; and visual impact.[3,5,22]

(b) Gas - North Sea

Resource Life

Known reserves of natural gas, at current consumption levels, may last for around 60 years.[35] Some commentators believe that, given the present growth in discoveries, gas will be available for the next 100 years and that "No real crisis in supply is imminent".[36] Others argue that the current growth in the use of natural gas will just about balance out the expected new discoveries, leaving a resource life still around the 60 year mark.[35] Natural gas is being used at increasing rates, because it is currently very cheap, and is seen as a 'clean' fuel.

Global Warming

Burning of natural gas creates the least amount of CO_2 per unit of heat than for any other fossil fuel, but it is still a considerable amount - around 60 kg.CO_2/GJ.[11] Moreover, natural gas (methane) is itself a potent greenhouse gas, so leaks in the system anywhere from the drilling rig to the domestic piping also contribute directly to global warming.[3,22]

Acid Rain

Natural gas contributes very little to acid rain: SO_2 emissions are virtually zero; and NOx emissions are very small compared to other fossil fuels.[34]

Emissions of the **Photochemical Smog** gases, **Particulates** and **Toxics** are likewise low from the combustion of natural gas.[34]

Risks

The chief risks associated with gas use are explosions (either at point of use or production), and storm or diving accidents at off-shore production platforms.[33]

Other

The local impacts of gas extraction are similar to those for oil extraction, with pipelines being the major source.

(c) LPG

LPG or bulk propane is a by-product of oil refining,[17] thus its impact rating for extraction is the same as that of oil (see below). On combustion, LPG is almost as clean as methane (natural gas), but gives off slightly more CO_2.[6]

(d) Oil

Resource Life

Known oil reserves will last only about 30 years[33] with expected discoveries increasing this figure by perhaps another 10 years.[35] The most optimistic estimate is for about 70-80 years[1] before it is all gone, but prices will surely rise long before then to make it much less of an attractive economic proposition for 'low-grade' use such as heating. UK production from the North Sea has already peaked, and has been declining since 1985/6.[36]

Global Warming

The emissions of CO_2 from oil combustion lie between those of coal and gas at around 80 kg.CO_2/GJ.[11] Methane is often also released or flared off during oil extraction, further contributing to the greenhouse effect.

Likewise with emissions causing **Acid Rain** and **Photochemical Smog**, and with **Particulates** and **Toxics** emissions, oil combustion falls between coal and gas.[34,3,22] Some further toxic hydrocarbons, such as benzene, are also emitted during the oil extraction process.[3]

Risks

Occupational risks in oil production are mainly concerned with safety on oil rigs, including blowouts and fires (e.g. Piper Alpha).

Other

Leaks and spillages of oil, either routine or accidental, also cause concern, as does the impact of pipelines and refineries.[3]

(c) Electricity (national grid)

In the UK, energy distributed over the national grid as electricity comes from three main sources:

Energy Sources of UK electricity supply[22]

Coal	68%
Nuclear	21%
Oil	8%
Hydro	2%
Other	1%

Since these figures were compiled, both natural gas and wind power have begun to be used more for generation, but they still only form a tiny fraction. The conversion of heat, from combustion or nuclear reactions, via steam, into electricity, is governed by the physical laws of thermodynamics, which limit the efficiency with which this can be performed. The best power stations achieve perhaps 40% conversion efficiency (e.g. 1 GJ of coal producing 0.4 GJ of electrical energy), whilst the average for the UK is probably around 33%.[1]

We have used these figures in combining the relative impacts of coal and oil combustion (see above) to arrive at an estimate for the impact of electricity generation, to which is also added that for the nuclear fuel cycle.

(f) Nuclear

Resource Life

Uranium is a very widespread element in the earth's crust, but only at tiny concentrations. Thus the level of economically extractable reserves depends crucially on price. At a price up to $130/kg (it is currently only $21/kg), reserves will last at 1980's production levels for 80 years.[36]

The contribution of the nuclear fuel cycle to Global Warming, Acid Rain, Photochemical Smog and Particulates is relatively slight, arising from the fossil fuel energy used in mining, refining and transporting uranium.[33]

Risks

The major concern with the nuclear industry is with the risks of: the decommissioning of old plant and the treatment or storage of highly radioactive wastes; a major accident (e.g. Chernobyl); long-term exposure to low levels of radiation; and security and defence issues.[3]

The long-term doses of low-level radiation from nuclear power plants are usually too small to measure against normal background radiation, and are normally derived by calculation instead.[33] Epidemiological studies of local populations have come up with conflicting results as to whether there is any real health risk from such exposures. However, the risks of major accidents, and the geological time spans for waste-containment, have deterred the markets from investing in nuclear power in recent privatisations.

Other

Uranium mining has significant local impact, and can cause health problems for workers.[33]

"No-one on the mains can say their electricity is entirely wind powered (or entirely wind free), no matter where they live. But the more wind farms are built, the less fossil fuel will be burnt and the less nuclear power stations will have to be built or kept going."

(Brian Horne[31])

4.4.2 Renewables

(a) Solar & locally produced wind/hydro/etc.

Toxics

Photovoltaic panels (used to generate electricity directly from sunlight) are manufactured in a process similar to that used in the semiconductor industry. Although the basic raw material is silicon, obtained from sand, a range of toxic chemicals is also used.[30,33]

Risks

Some campaigners against windfarms have drawn attention to the risks of propellor blades flying off fast spinning wind turbines. So far this is not known to have caused injury to anyone.[32]

Other

Both noise and visual intrusion have been cited against wind-powered generators. Others consider them to be objects of great beauty. Sensitive siting may be called for.[7]

Alert

It is possible that some evacuated tube solar collectors (high efficiency solar panels for heating water) may contain ozone-depleting gases. Check with the manufacturer.[6]

(b) Wood & other biomass, biogas etc

Resource Life

Wood and other biologically produced raw materials, collectively known as 'biomass', may be either sustainably produced in well-managed systems, or may come from the destructive over-harvesting (effectively mining) of natural resources. This analyis assumes that anyone choosing a biomass fuel for ecological reasons will go to the trouble of obtaining it from a well-managed source. Any energy crop depends on the health and fertility of the soil it is grown on, and intensive, large scale, high chemical input farming "will have difficulty maintaining any environmental credibility."[26] Burning of crop residues such as straw is also a biomass option, but again, in order to maintain the health of the soil, there is a limit on how much can be taken and not returned to the soil before it loses structure and fertility. US studies suggest a maximum of 35% of residues.[34]

Global Warming

It is commonly assumed that for biomass fuels the emissions of CO_2 from combustion are balanced out by the absorption of CO_2 during plant growth. Thus the combustion of biomass is CO_2 (and therefore greenhouse) neutral, but only so long as the whole cycle of planting, growth, harvesting and replanting (or regrowth with coppicing), is in place.[17,34]

With systems such as biogas, where wastes are digested by bacteria to produce methane, leaks of gas will contribute to global warming.

Air pollution, from emissions that contribute to **Acid**

> *"The long-term availability of usable sources of energy and our ability to use them wisely are the factors which will be predominant in deciding the length and nature of the human race's inhabitation of our planet... "*
>
> *(Andrew Porteus[1])*

Rain and **Photochemical Smog**, as well as **Particulates** and **Toxics** emissions, are the major concern with the burning of biomass fuels. The actual amounts and types of pollutants vary widely, depending on the type of fuel, its state (wet or dry, fresh or decomposed etc.) and the burning conditions. Biogas is essentially methane, the same as natural gas, but other impurities may be present. Well-dried wood, burnt in a modern stove that allows for a secondary combustion air-flow, can be a relatively clean option.[17,36] Stoves are now even available with catalytic converters.[37] Scrap wood that has been treated with preservatives may also release toxic chemicals such as arsenic or dioxin on burning.

Larger scale systems, whether burning wood chips or biogas, suited to larger institutions or district heating systems, tend to be less polluting.[26]

Risks

The inefficient burning of biomass fuels for indoor cooking in developing countries is a major source of health problems.[34]

The use of open fires in the home also increases the risk of house fires,[33] and the storage of biogas (methane) may pose a risk of explosion.

Other

The use of land specifically for the growing of energy crops displaces other uses such as food crops.[34]

Large-scale biomass treatment plants may create water pollution problems.[34]

Did You Know?

The Energy Act 1976 - Statutory instrument 1980/1013: Fuel and Electricity (Heating) (Control) (Amendment) 1980 stipulates a maximum temperature for non-domestic heated buildings of 19°C - yet current practice ignores this, usually maintaining temperatures greater then 20°C.[4]

4.5 Alternatives

The following section looks at the practicalities and economics of the renewable energy alternatives. For further information on these, the first port of call has to be the Centre for Alternative Technology at Machynlleth in Wales. Their bookshop on its own makes travelling to the Centre worthwhile, never mind the practical displays themselves, but it also does an excellent mail order service. CAT also publishes extensive lists of suppliers of equipment for utilising renewable energy as well as consultants and other useful contacts.

The Centre for Alternative Technology, Machynlleth, Powys SY20 9AZ tel: 01654-702400 fax: 01654-702782 bookshop: 01654-703409.

Basic Passive Solar Principles

Overhangs protect windows from high summer sun

Heat from sunlight absorbed during the day, and released during the night

Roof lights etc. admit light & heat to rear of building

Thermally massive construction of walls, floor etc. (e.g. earth, brick, concrete)

Winter sun enters windows at a lower angle

Insulation on outside of thermal mass

Large glazed areas facing South

Little or no glazing to the North

4.5.1 Passive Solar

The Building Research Establishment estimates that natural solar gains, mostly unintentional, already contribute around 15% of the heating energy used in UK homes.[11] Passive solar architecture aims to increase this contribution by the use of large areas of south-facing glazing, and advocates believe that, combined with high levels of insulation, thermally massive construction, and the additional gains from cooking, lighting and the occupants themselves, no additional space heating energy is needed. Indeed, care must be taken to avoid overheating in summer, and adequate ventilation and humidity reduction become the key issues.

In less than optimal settings, such as retrofitting existing buildings, or where site restrictions do not permit perfect solar orientation, backup heating may be required. The key to efficiency in this case is to use a flexible system with fast response times, so that the solar component is allowed to contribute useful heat whenever it can.

Passive solar is simple in concept and is gaining widespread popularity. It entails little or no extra capital costs when included in the design of new building unlike active solar techniques. There are numerous books etc. on the subject, and many different ways of designing for passive solar gain. (The pictures here just show the very basic ideas.) A growing number of examples of successful buildings exist around the country. Well-known examples include Looe Junior and Infants School in Cornwall, and many new homes in Milton Keynes.

In order to promote passive solar building, Friends of the Earth want the Building Regulations to provide example designs with passive solar features that are deemed to conform to the regulations. This would be useful for small builders etc. without facilities for the more complex computer modelling that some designers use. FoE also want Building or Planning Regulations to include measures to prevent the overshadowing of neighbours and thereby reducing their solar performance. They estimate that at least 3 million passive solar buildings (new and retrofit) could be in place in the UK by 2020.[7]

Passive solar design also makes good economic sense.

The additional cost for glazing, super-insulation etc. is estimated to be no more than 5-10% of the cost of a conventional new building. The energy savings of such a design will be greater than the extra payments on an ordinary mortgage. "This fact [the economic good sense of passive solar design] has not yet got through to builders or mortgage lenders to the domestic market, but many commercial buildings incorporate atria and other passive solar features for the benefits they bring to space heating and daylighting."[36]

4.5.2 Active Solar

(a) Domestic Hot Water

The use of solar panels on the roofs of buildings to provide hot water is now a fairly common sight. Solar systems integrated into the roof fabric can be more economical and look better too. The technology ranges from sophisticated (and expensive) evacuated tube collectors to black-painted radiator panels. Solar collectors are usually able to supply all domestic hot water needs for only part of the year. Additional heating will be needed

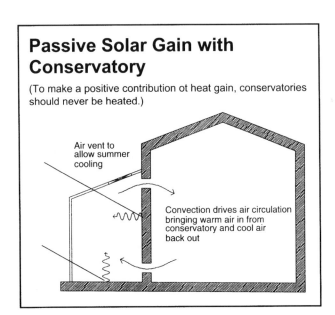

Passive Solar Gain with Conservatory

(To make a positive contribution ot heat gain, conservatories should never be heated.)

Air vent to allow summer cooling

Convection drives air circulation bringing warm air in from conservatory and cool air back out

for the winter months.

According to the Centre for Alternative Technology, a 'typical' solar system might provide around 40-60% of the hot water for a household, saving around 1,500 kWh a year.[28] At a capital cost of between £1,500 and £3,500 for a professional installation, this is not very economic in conventional terms - it would take a long time to pay back such costs compared to the price of equivalent fossil-fuelled heat. Given that an additional boiler would be required to supply the shortfall, the solar collector may save as little as £25 per year (compared with a high-efficiency condensing gas boiler). The savings in pollution are not taken into account in such calculations. Cheaper and simpler 'home-made' systems on the other hand, (such as under-slate micro-bore piping) although less thermally efficient, may well be more economic.[28]

(b) Space Heating

In Britain, interseasonal heat storage is essential if solar energy is to provide space heating in the winter. This is most often achieved with the use of large, well-insulated and often underground, water tanks. These are heated up by solar panels over the summer months, and then the heat is released gradually over the winter. Heating systems based on interseasonal storage need to be designed specifically to work with relatively low temperatures of hot water - under-floor heating is well suited,[29] as is warm-air circulation.[10]

Experience of interseasonal heat stores at the Centre for Alternative Technology shows that they are practical (their system provided 80% of space heating requirements, and could easily have been better), but not particularly cost-effective on a small scale.[29] However, on a larger scale it looks much more promising. With larger systems, the efficiency of the collectors, and heat loss from the store are much less crucial. There are successful examples of such systems in Sweden, such as at Lambohov, where a group of 55 houses gets 90% of their winter space heating from 2,500m² of roof collectors plus a 10,000m³ store.[29]

Developments in thermal storage are continuing, with one promising technology using the phase-change (freezing and unfreezing) effects of various salt solutions, and it is likely that storage will become economic in the longer term. It may therefore be a good strategy to include *space* for such a heat store in buildings designed now, even if this space is only used as a wood store for now.[41]

(c) Photovoltaics

Photovoltaic (PV) panels are semiconductor devices that convert sunlight directly into electricity. Given their high cost and the difficulty of storing electrical energy efficiently, they are not normally considered suitable for providing energy for heating loads.

A small-scale PV system, for example, might have a capital cost of around £5,000 for a small domestic caravan supply, designed to give under 1 kWh per day - hardly enough for heating. The price of PV panels is still falling, but they only form around 1/3 of the total capital cost when system includes battery storage.[30]

However, where peak electrical loads coincide with brightest sunshine, so that storage is not an issue, PV technology is more promising. Such a situation occurs with (air-conditioned) office buildings, and the use of PV cladding panels to high-tech offices is becoming popular - especially as PV panels are no more expensive than marble or granite cladding, at around £800 - £1,000/m².[30,36] A 40 kW system is in use on the southern face of a building at the University of Northumbria.

Photovoltaics is a technology to watch for the future. As costs come down, and PV panels are integrated into roof tiles etc. it will become more economical, even becoming economic to sell energy back to the grid for 'storage'.[36]

4.5.3 Biogas

Waste matter from farm animals or sewage is the most common raw material for biogas digesters, which biologically break down the raw material to produce methane gas and a valuable fertiliser residue. Such systems are widely promoted in developing countries, especially China, and on a small scale, are only really suited to a rural agricultural setting. (Alternatively, municipal refuse sites are a good source of biogas, but these are not strictly a renewable resource.)

TV's 'The Good Life' was hopelessly unrealistic in portraying the waste from two people and one pig as providing a useful amount of fuel for a generator. Between 4 and 8

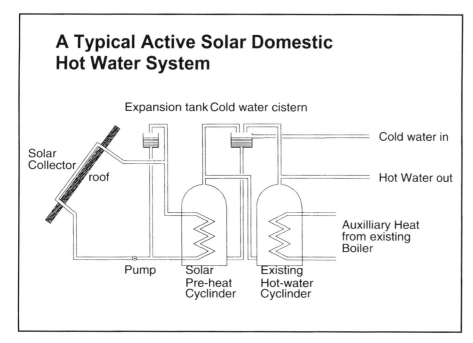

A Typical Active Solar Domestic Hot Water System

Expansion tank Cold water cistern

Solar Collector / roof

Cold water in

Hot Water out

Auxilliary Heat from existing Boiler

Pump   Solar Pre-heat Cyclinder   Existing Hot-water Cyclinder

large farm animals would be needed to provide enough biogas energy for the cooking energy alone of one family.[27]

However, for those with a good supply of raw material, such as the slurry from a dairy farm, biogas can be very economical. The investment in plant can be paid back in under eight years compared to bought-in gas,[1] and a waste disposal problem has been turned into valuable fertiliser.

In cold climates, the economics of biogas suffer, as a further heat source (usually a fraction of the gas produced) is needed to keep the digester at an optimal temperature. It is therefore least efficient in winter when it is needed the most.[27]

Biogas is most suited to moderately large-scale operations where the potential is considerable, but allowance needs to be made for management and maintenance costs, as well as the long-term regularity of biomass supply.[36]

4.5.4 Wood

Wood burning stoves are now available that, with well-dried fuel and some attention, will burn relatively cleanly.[18] The justification for counting wood as a renewable fuel depends upon the source being from a well-managed woodland or coppice with sufficient replanting to maintain the supply. Sometimes scrap or waste timber is counted as a renewable source too.

Either way, if a wood-fuelled heating system is being installed, the responsible designer will also consider the specific supply that will be used. It is also worth bearing in mind that, like all solid fuel heating, using wood requires some input of work from the user - if not actually chopping and stacking wood, then at least in tending the stove. This may or may not be a chore, depending on attitudes to work, but the user is more aware of the amount of energy used if it has to be manually carried rather than arriving down a pipe.

Short-rotation coppicing is currently a popular option. Many broadleaf or deciduous trees will sprout new growth when cut down to a stump, the new growth being very vigorous. Short rotation coppicing is often done on a 7-year basis, where one seventh of the total area coppiced is cut each year, or sometimes even shorter rotations of 2-3 years are used. Coppicing is suited to a range of scales from single homes to large commercial operations supplying a district heating system.[26]

The proviso is of course that enough land is available. A hectare (2.47 acres) of typical arable coppice might yield between 8 and 20 dry tonnes per year, which when burnt would release between 100 and 260 GJ of heat.[42]

A good modern wood stove with back-boiler might be 65% efficient.[26] The importance of insulating well and maximising solar gains is obvious in reducing the amount of land needed.

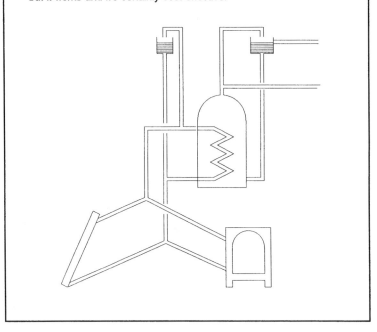

A Simple System Combining Solar Domestic Hot Water with a Wood-burning Stove

This sytem is used by CAT. The solar panel provides hot water on warm sunny days; the wood stove provides both space and water heating on cold days. "It may not technically be the most efficient system we have, but it works and it's certainly cost effective."[28]

Additional Benefits

The growing of trees is popular with almost everyone, for its environmental and aesthetic benefits. Coppicing may provide rural employment, and has particular wildlife benefits different to normal woodland. It may also require less management than conventional forestry as replanting is seldom needed.[38] Wood ash, unlike coal ash, may be used on the garden as a fertiliser, and contains useful amounts of potassium, an element missing from many other organic fertilisers.[39]

4.5.5 Wind Power

Wind power is fun, but not practical in most situations, and is unlikely to perform well in built-up areas due to low wind speeds, turbulence etc.[31] Turbines can also be dangerous in such situations - though vortex systems (under development) seem promising in both respects. Wind turbines come in all sizes from tiny ones that keep boat and caravan batteries charged to the huge ones that adorn the hill-top wind-farms of Wales.

To use wind-generated electricity directly for water heating via an immersion heater (e.g. a 3kW load for 4 hours a day) would need, just for this and no other loads, a 10kW (7m diameter rotor) machine costing £10,000 in a good inland location. On a very windy site you could maybe get away with a smaller one, perhaps 1.5 kW, 3m diameter, costing £3,000.[32] In either case, supply would still depend on the wind blowing enough at the right time.

However, for a house in a very windy spot, with no mains electricity, a wind turbine plus battery system may be the cheapest way to power a few lights, a radio and small fridge etc. - at a capital cost of around £2-3000 (or cheaper if DIY). Larger systems, with a back up diesel generator (5kw), would cost at least £10,000, and would do a bit of water heating too when it was very windy and the batteries were already charged.[31]

On a larger scale, windpower can be economical in the right location. The Scottish island community of Foula, for example, has a wind generator plus diesel back-up and hydroelectric storage system that cost about £500,000 to install. Previously with no mains electricity at all, 45 households now have electricity "coming out of their ears" in winter, and though it can be limited at other times, and is therefore rather different to mains supply, the users are satisfied.[31]

4.5.6 Hydro Power

Hydro power is obviously only suited to a very few locations, and is most likely be found in rainy, hilly country. There are 80,000 mini hydro plants in China. Capital costs are likely to be in the range of £1,000 to £2,000 per kW for small installations of around 10-30 kW size (1983 prices).[24] However, where it is suitable, hydro power is the most likely to provide a consistent flow of energy when it is most needed during the winter. "Wherever transmission lines can furnish unlimited amounts of reasonably priced electric current, it is usually uneconomical to develop small and medium-sized [water-power] sites."[25]

Combined Heat & Power (CHP)

Combined Heat & Power (CHP) is a technique for making the most out of any given fuel. Whilst CHP is a much more efficient way to utilise fuels, it is still in the main designed for fossil fuels such as oil, gas or coal, although it could technically be powered by biomass fuels.

The fuel is first burnt, in anything from a tiny diesel generator to a full scale 'combined cycle' gas and steam turbine generator set, to produce electricity. The left-over heat, usually in the form of hot water or steam, is then used directly as well, for heating buildings or for other processes. The overall fuel efficiency can be as high as 85%, compared to the 35% of a typical coal-fired power station.[12]

Small and micro CHP units are now available, with an electrical output up to 160kW, and a thermal output 2-3 times that figure, suitable for use in housing schemes or blocks of flats either new-build or retrofit, in hotels, hospitals, residential homes, swimming pools and similar-sized projects.[5,14]

The smaller CHP systems make best economic sense where demand for electrical power coincides in time with demand for heating, as there is no practical way to store electrical power at this scale. The 1983 Energy Act requires electricity boards to buy surplus electricity from CHP schemes.

KEEP WARM THIS WINTER...

AND NO-ONE IS SEARCHING HARDER THAN B.N.F.L. FOR NEW WAYS TO PROTECT AND PRESERVE OUR NATURAL ENVIRONMENT. TO FIND OUT MORE, CALL ~

...WATCH LOTS OF ADS ON THE T.V.

The Embodied Energy of a Wind Turbine and a Stove

A European study, that looked at a whole range of different wind turbines, found that most machines had generated as much power as went in to their manufacture in less than half a year. For the rest of their life (expected to be around 20 years) they were in energy profit,[40] and therefore generated, at the least, 40 times their embodied energy.

A stove or boiler uses fuel rather than generating power, so will not make an energy profit. The following example has been calculated for a typical wood-burning stove. It demonstrates in just how short a time a stove or other heating appliance will use up the same amount of energy as went into its manufacture. (About 12 days with 8 hours use a day.)

Stove:	a Coalbrookdale 'Much Wenlock' stove with boiler, rear outlet
Output:	3.9 kW space heating, 7.3 kW water heating
Weight:	184 kg.
Embodied Energy of Cast Iron:	32 GJ/tonne
Assuming stove is all cast iron, embodied energy of stove:	0.184 x 32 = 5.888 GJ.
Converting Units:	5.888 GJ = 5.888 / 0.0036 = 1,636 kWh
Stove Output:	3.9 + 7.3 = 11.2 kW
Estimated Efficiency:	65%
Energy Input for 11.2 kW Output:	11.2 / 0.65 = 17.2 kW
Time to Burn Equivalent of Embodied Energy of Stove:	1,636 kWh / 17.2 kW = 95.1 hours
If stove lasts 20 years, how much energy it burns in its life:	20 x 2912 (hours in a year @ 8 hours a day) x 17.2 kW = 1,001,728 kWh
How many more times the energy it burns in its life is compared to its embodied energy:	1,001,728 kWh / 1,636kWh = 612 times

Greenie Points

BREEAM/New Homes version 3/91 awards credits on a 6-point scale based on the calculated CO_2 production from fuel use per square metre of floor area.[15] Their scale gives a useful indication of the amounts of energy used by a typical house, which we have calculated for the last column in this table.

BREEAM Credit	CO2 emissions (kg/m2.yr)	Example House	Energy Use of Example House (100 m2 floor area) (GJ/yr)
1	105-91	Built to 1990 Building Regs insulation requirements, plus space heating by conventional modern well-controlled gas boiler	181 - 157
2	90-71	Super-insulation plus all electric heating;OR Building Regs insulation, space & water heating by conventional gas boiler	156 - 123
3	70-56	Building Regs insulation plus space & water heating by modern gas boiler & cooking by gas; OR super-insulation plus heating by condensing gas boiler	122 - 97
4	55-46	Building Regs insulation but plus double glazing with low-E glass, additional loft and door insulation plus space & water heating by condensing gas boiler & cooking by gas	96 - 79
5	45-36	Super-insulation plus space & water heating by condensing gas boiler & cooking by gas	78 - 62
6	<36	Super-insulation plus heating by condensing gas boiler, high efficiency heat exchanger (heat recovery coil) for hot water and low-temperature heat distribution system.	< 62

4.6 References

1. Dictionary of Environmental Science & Technology (Andrew Porteus, John Wiley & Sons, Chichester, 1992)
2. Environmental Chemistry 2nd Edition (Peter O'Neill, Chapman & Hall, London, 1993)
3. The World Environment 1972-1992 - Two Decades of Challenge (M. K. Tolba & O. A. El-Kholy (eds), Chapman & Hall, London, for The United Nations Environment Programme, 1992)
4. The Green Construction Handbook - Going Green (J T Design & Build, 1993)
5. Green Design: A Guide to the Environmental Impact of Building Materials (Avril Fox & Robin Murrell, Architecture Design & Technology Press, London, 1989)
6. Greener Building: Products & Services Directory (Keith Hall & Peter Warm, Association for Environment Conscious Building, Coaley,)
7. Energy Without End, M. Flood (Friends of the Earth, London 1991)
8. BRE Information Paper IP 16/90, The BRE low-energy office: an assessment of electric heating.
9. BRE Information Paper IP 6/91, Condensing Boilers: how they compare with other systems in the BRE low-energy office
10. BRE Information Paper IP 1/89, Domestic warm-air heating systems using low-grade heat sources
11. Domestic Energy Fact File (Shorrock, Henderson & Brown, BRE, 1992, updated 1993)
12. Energy Use in buildings and CO2 emissions (Shorrock & Henderson, BRE, 1990)
13. Memorandum from the Centre for Alternative Technology to the Commons Select Committee on Energy (3rd October 1991)
14. Buildings & Health - The Rosehaugh Guide to the Design, Construction, Use & Management of Buildings (Curwell, March & Venables, RIBA Publications, 1990)
15. BREEAM/New Homes version 3/91 (BRE, 1991)
16. C for Chemicals - Chemical Hazards and How to Avoid Them (M. Birkin & B. Price, Green Print, 1989)
17. Out of the Woods - Environmental Timber Frame Design for Self Build (Pat Borer & Cindy Harris, Centre for Alternative Technology Publications, Machynlleth, 1994)
18. Eco-Renovation: The ecological home improvement guide (Edward Harland, Green Books, Dartington, 1993)
19. Energy Efficient Building: A Design Guide (Susan Roaf and Mary Hancock, Blackwell Scientific, Oxford, 1992)
20. Response to DoE's 'Climate Change - Our National Programme for CO2 Emissions' (Friends of the Earth)
21. Back From the Brink - Greenhouse Gas Targets for a Sustainable World (Jaqueline H W Karas, Friends of the Earth, London, 1991)
22. The UK Environment (Department of the Environment, HMSO, London, 1992)
23. Environmental Building B4 (Pat Borer, Centre for Alternative Technology, Machynlleth, 1994)
24. Micro Hydro Electric Power, Technical Papers 1 (Ray Holland, Intermediate Technology Development Group 1983)
25. Low Cost Development of Water Power Sites (Hans W. Hamm, VITA (Volunteers in Technical Assistance, Arlington, Virginia)
26. What Are Biofuels? (Centre for Alternative Technology Factsheet R5, CAT, Machynlleth)
27. Running a Biogas Programme: A Handbook (David Fulford, Intermediate Technology Publications, London 1988)
28. Tapping the Sun: A Solar Water Heating Guide (Brian Horne, Centre for Alternative Technology, Machynlleth 1992)
29. A Solar Heating System with interseasonal storage (R W Todd, Centre for Alternative Technology, Machynlleth 1995)
30. Wired Up to the Sun (Paul Allen, Centre for Alternative Technology, Machynlleth 1994)
31. Where The Wind Blows (Brian Home, Centre for Alternative Technology, Machynlleth 1993/4)
32. Choosing Windpower: A Guide to Designing Windpower Systems (Hugh Piggott, Centre for Alternative Technology, Machynlleth 1991)
33. WHO Commission on Health and Environment, Report of the Panel on Energy (World Health Organisation, Geneva 1992)
34. Green Energy: Biomass fuels and the environment (United Nations Environment Programme 1991)
35. Beyond the Limits: Global Collapse or a Sustainable Future (Meadows, Meadows & Randers, Earthscan Publications, London 1992)
36. The Future of Energy Use (R Hill, P O'Keefe & C Snape, Earthscan Publications, London 1995)
37. EFEL Stoves - Distributors' catalogue (Euroheat Ltd, Worcestershire)
38. Forestry Practice (Edited by B G Hibberd / Forestry Commission, HMSO London, 1991)
39. Organic Gardening, DK Pocket Encyclopaedia (Ed. Geoff Hamilton, Dorling Kindersley, London 1991)
40. How Renewable is Wind Energy? (Special Feature in Windirections - magazine of the British Wind Energy Association, J Schmid, H P Klein & G Hagedorn)
41. Design with energy: the conservation and use of energy in buildings (J Littler & R Thomas, Cambridge University Press, Cambridge 1984)
42. Willow as a sustainable resource (Halliday, Smedon & Howes, CIBSE Conference 1995)

THE GODS OF ALTERNATIVE ENERGY...

Polyp

Thermal Insulation Materials   5

5.1 Scope of this Chapter

This chapter looks at the main materials available in the market for use as building insulation. Though there are a wide variety of insulation materials available, they are not all suitable to every application. For instance, a brick faced cavity wall may well let in a fair amount of moisture, and so any material used in this situation must be able to withstand the effects of this. The choice of construction method may thus determine which insulants are chosen - or, indeed, vice versa.

There are also a whole range of composite products available, usually as boards and panels, combining layers of such materials as plastic foams and mineral fibres with stiffer facings. We couldn't cover the whole variety here, but hope to have covered the main constituents.

5.2 Introduction

If this chapter were being written in the late 1980s, ozone-depletion would have been the big issue. CFCs were then the major blowing agents used in plastic foam insulation. Some of the more forward-looking manufacturers were advertising their wares as 'CFC-free' or 'ozone-friendly'. And just about any insulation, so long as it wasn't a foam blown with CFCs, was considered acceptably green.

Things have moved on a lot since those times. International agreements brought about an end to CFC production by the end of 1995. Already most foam manufacturers will have stopped using these gases.

Other chemicals known as HCFCs, have been developed as 'interim measures' - but their ozone-depleting potential is still significant, especially in the short term. (Estimates of this vary from between 5% and 20% that of CFCs.) These gases are also regulated by phase-out agreements, but will still be allowed for a number of years.

5.2.1 HCFCs: Out of the Frying Pan - Into the Fire?

Chemical manufacturers have come up with another 'alternative' to CFCs, known as HFCs. These don't have the chlorine content of CFCs, and so do no harm to the ozone layer, and are not regulated by the ozone agreements. However, HFC gases have very high global warming potential, about 3,200 times that of carbon dioxide. The British scientist, Joe Farman, credited with discovering the Antarctic ozone hole, aptly described the move from CFCs to HFCs as being "out of the frying pan, into the fire".

There are already a number of alternative gases for foam blowing, such as CO_2 or the hydrocarbons (e.g. pentane). These are widely available (i.e. un-patentable) and cheap, do not harm the ozone-layer, and have a relatively low global warming effect. Given that these alternatives exist, building designers wanting to use blown plastic foams, and also wishing to ensure that their designs have minimal impact on the ozone layer and on global warming could therefore take care to specify materials that are CFC-, HCFC-, and HFC-free.

Environmental concern has always been more widely focused than on just the ozone layer. And since CFC emissions appear to have (at least partially) been addressed by intergovernmental agreement, attention now falls on a wider range of environmental impacts.

5.2.2 Thermal Conductivity

Thermal conductivity, commonly known as the K-value, is a measurement of how much heat will move through a given amount of a material. For insulation materials therefore, a low K-value is the aim. A material with a higher K-value will require to be used in a thicker layer to achieve the same degree of insulation as a material with a lower value. The following table lists the values for a range of materials in ascending order of K.

(Source: 8)

Material	K Value (W/mK)
Polyurethane Foam	0.024-0.039
Rock Wool	0.03-0.04
Glass Wool	0.032-0.04
Polystyrene Foam	0.033-0.035
Phenolic Foam	0.036
Wool	0.037
Cellulose Fibres	0.037
Urea-formaldehyde foam	0.038
Corkboard	0.040
Vermiculite (expanded)	0.047-0.058
Foamed Glass	0.050-0.052
Softboard	0.055
Wood-wool slabs	0.093
Compressed Straw Slabs	0.101

Some Other Materials for Comparison

Thatch	0.072
Timber (pine)	0.138
Diatomaceous Earth Brick	0.141
Aerated Concrete	0.18
Glass	1.05
Brickwork (common)	1.15
Stone (granite)	2.9
Copper	400

'Greenie Points'

The Canadian ecolabelling scheme 'Environmental Choice' has awarded a label to insulation made from recycled wood-based cellulose.

The Japanese Ecomark is awarded to all thermal insulation.[17]

	£	Production									Use				
	Unit Price Multiplier	Energy Use	Resource Depletion (bio)	Resource Depletion (non-bio)	Global Warming	Ozone Depletion	Toxics	Acid Rain	Photochemical Oxidants	Other	Energy Use	Durability/Maintenance	Recycling/Reuse/Disposal	Health	Other
Insulation Materials															
Cellulose Fibres	n/a	•												?	
Compressed Straw Slabs	n/a	•	•										•		
Cork	7.2	•											•		
Foamed Glass	16.7	●		•			●	●	●	●					
Glass Wool	1.0	●		●			●	●	●	●			●	•	
Phenolic Foams	n/a	●		•	?	?	●	●	●	●				●	HFCS, HCFCS
Polystyrene - Expanded	3.1	●		•			●	●	●	●				•	
Polystyrene - Extruded	8.2	●		•	?	?	●	●	●	●				•	HFCS, HCFCS
Rigid Urethane Foam	4.9	●		•	?	?	●	●	●	●				●	HFCS, HCFCS
Rock Wool	1.0	•		●			●	●	●	●			●	•	
Softboard	9.5	•	•										•		
Softboard + Bitumen	8.7	•	•	•			•	•	•	•			•		
Urea-Formaldehyde Foam	n/a	●		•			●	●	●	●				●	
Vermiculite (Expanded)	n/a	•		•						●				?	
Wood-Wool Slabs	11.8	●	•	•	•		•	•	•	•			•		
Wool	10.4	•													

5.3 Best Buys

Best Buy:	Wool, Cellulose Fibre, Cork, Strawboard, Softboard, Wood-wool
Second Choice:	Foamed Glass

The insulation products that come out with the smallest environmental impact on our Table are those most closely associated with 'natural' products: wool, cellulose, cork, strawboard, softboard and wood-wool, and these are our Best Buys.

The first three of these are thermally as good as more conventional insulators. Only wool is suitable however for use in moist situations such as in brick cavity walls. The others are well suited to breathing timber-frame construction or roof cavities. Warmcel, the only product in this report marketed as 'made from recycled materials', appears on the table and in this report as 'Cellulose Fibres'

Of the more 'conventional' insulation materials, foamed glass comes out well (partly because it incorporates no petrochemical based binders or preservatives). It performs well in the wet and is perfect for some situations, but is expensive.

Slag-wool (see Alternatives, p49), if it were available in the UK, would probably be a best buy of the mineral wools, but there is currently little to choose between glass-wool and rock-wool.

5.7 References

1. Green Design: A Guide to the Environmental Impact of Building Materials (Avril Fox & Robin Murrell, Architecture Design & Technology Press, London, 1989)
2. Rachel's Hazardous Waste News #367, December 9, 1993 (Environmental Research Foundation, Annapolis, MD USA)
3. Secretary of State's Guidance - Exfoliation of vermiculite and expansion of perlite (EPA 1990 part 1) (HMSO, London, 1991)
4. The Building Regulations 1991, Approved Document D, Toxic Substances (Department of the Environment, 1991)
5. Hazardous Building Materials (S R Curwell & C G Mach, E & F N Spon Ltd, 1986)
6. Construction Materials Reference Book (ed. D. K. Doran, Butterworth-Heinemann, Oxford, 1992)
7. Buildings & Health - The Rosehaugh Guide to the Design, Construction, Use & Management of Buildings (Curwell, March & Venables, RIBA Publications, 1990)
8. Mitchell's Materials 5th Edition (Alan Everett, Revised C M H Barritt, Longman Scientific & Technical, Harlow, 1994)
9. Eco-Renovation: The ecological home improvement guide (Edward Harland, Green Books, Dartington, 1993)
10. Energy Efficient Building: A Design Guide (Susan Roaf and Mary Hancock, Blackwell Scientific, Oxford, 1992)
11. Greener Building: Products & Services Directory (Keith Hall & Peter Warm, Association for Environment Conscious Building, Coaley 1992)
12. Building For A Future Winter 1994/5 vol.4 no.4 (Association for Environment Conscious Building, Coaley, 1995)
13. Out of the Woods - Environmental Timber Frame Design for Self Build (Pat Borer & Cindy Harris, Centre for Alternative Technology Publications, Machynlleth, 1994)
14. Dictionary of Environmental Science & Technology (Andrew Porteus, John Wiley & Sons, Chichester, 1992)
15. Environmental Building B4 (Pat Borer, Centre for Alternative Technology, Machynlleth, 1994)
16. The Ecological and Energy Balance of Foamglas Insulation (Pittsburgh Corning (UK) Ltd, Reading, 1992)
17. Ecolabelling of Building Materials and Building Products, IP 11/93 (C J Atkinson & R N Butlin, Building Research Establishment, Garston, 1993)
18. The World Environment 1972-1992 - Two Decades of Challenge (M. K. Tolba & O. A. El-Kholy (eds), Chapman & Hall, London, for The United Nations Environment Programme, 1992)
19. Glass (excluding lead glass) manufacturing processes PG 3/3(91) (Department of the Environment, HMSO, London, 1991)
20. Glass Fibres & Non-asbestos Mineral Fibres IPR 3/4 (Her Majesty's Inspectorate of Pollution, HMSO, London, 1992)
21. Oil & Energy Trends Annual Statistical Review (Energy Economics Research, Reading, 1990)
22. CFCs and the building industry - IP 23/89 (D J G Butler, Building Research Establishment, Garston, 1989)
23. Fashion Victims. The Globe, No.26. Winter 1995/96
24. Wool Record, October 1995.
25. Scab Wars. Friends of the Earth 1993
26. Ecological Building Factpack. (R. Pocock & B. Gaylard) Tangent Design Books. 1992
27. Environmental Building News Vol. 5 Pt.1. Jan/Feb 1996
28. The Green Home Handbook. (G. Marlew & S. Silver). Fontana. 1991
29. PVC: Toxic Waste in Disguise. (S. Leubscher (Ed).) Greenpeace International, Amsterdam. 1992
30. Man Made Fibres - Assessment of Carcinogenicity. Answers to Ministers Questions, the Health Council. Publication 1995/18, The Hague, 8 September 1995.

Masonry Materials

6

6.1 Scope of this Chapter

This chapter looks at the environmental impact of the main options available in the market for building masonry walls. These include bricks and concrete blocks, both of which come in various forms, and also natural stone. The different options in the ingredients used for mortar are also looked at.

'Greenie Points'

The environmental assessment method BREEAM/New Homes Version 3/91, devised by the Building Research Establishment, awards 1 credit for specifying the majority (>50%) of masonry material in walls to be recycled or reused. e.g. re-used bricks etc., or cements, mortars, blocks etc. containing fly-ash or blastfurnace slag.[19] In Japan, cement with 50% blastfurnace slag is awarded the Japanese Ecomark.[15]

6.2 Introduction

Masonry products all have considerable environmental impact during their production. They are high volume, high density materials, made from quarried non-renewable resources, and mostly associated with high energy inputs. In contrast, once built, they are very low maintenance, and most are not reported to have any known hazards or impacts (except for lightweight aggregate blocks - see page 56).

The strength, durability and inert nature of bricks, blocks and stone are the qualities that make them so useful as building materials. These qualities also make them readily re-usable in their original form. There is a thriving market for reclaimed bricks and stone, coming mostly from demolished Victorian buildings. A number of complete new buildings have recently been built using entirely reclaimed masonry, and while pubs seem to be the most common users (perhaps for aesthetic rather than environmental reasons), new housing is also being built from reclaimed bricks. Examples in-clude the Brewers' Wharf pub at Merry Hill in the West Midlands, the Seabird pub in Bridlington, a pub at the Somerfield Centre in Wilmslow, and housing developments by Berkswell in Solihull and by Crosby Homes at Cropready, Banbury.

Due to the high density and the large quantities used of masonry materials, transport costs are significant. This is therefore one area of modern consumption that still has some local and regional variation in supply. Any decisions about materials choices should therefore take into account local availability.

New developments in greening the masonry materials industry seem in the main to be confined to making improvements to existing processes rather than developing new products or processes. With fuel costs the major part of brick and cement manufacturer's expenses, both these industries are continually looking for improvements in energy efficiency - but progress is slow. The use of alternative fuels, whether landfill methane or derived from toxic waste, is currently a popular 'environmental' option with manufacturers. But fuel substitution does not address the fundamental issue of energy consumption.

The use of industrial by-products is also hailed by some as an environmental improvement - such as the use of pulverised-fuel ash from power stations. As a short term expedient in reducing the demand for another manufactured product, using such by-products may be environmentally beneficial. But if the by-product is itself from an industry that is unsustainable, the use of that by-product in the long term creates reliance on, and economic support for, those industries, which makes the use of such by-products unsustainable. Questions have also been raised about the safety of using some industrial by-products in buildings (see Lightweight Blocks, page 56).

Looking to the future, the supply of reclaimed masonry is unlikely to be enough to meet demand, especially if more care is taken to re-use buildings rather than demolish them. The improvements in energy efficiency in firing bricks and in burning limestone for cement or lime are also likely to reach limits. Techniques such as building with earth will be one way forwards to a sustainable building industry. But there is also an urgent need for new sustainable alternatives to cement and lime.

6.3 Best Buys

The best overall environmental option for building masonry walls is to use reclaimed brick or stone, with a pure lime mortar.

Within each category of materials, best buys are as follows:

(a) Bricks

If reclaimed bricks are not suitable, then perforated ordinary clay bricks are best.

(b) Concrete Blocks

We have not come across a market for reclaimed concrete blocks. The best buy for concrete blocks depends upon the intended building's design. Ordinarily, dense blocks will be best, but if it is necessary that the block contribute significantly to insulation levels, then aerated blocks are the best buy. (See Embodied Energy below.) We do not recommend lightweight-aggregate blocks.

(c) Stone

Reclaimed stone is best, but if unsuitable or unavailable, newly quarried stone, especially if local, is a better buy than any of the manufactured options, though it may well be expensive.

(d) Mortars

A pure lime mortar, made with just sand and no additives is environmentally best. If this is impractical, lime mortars with pozzolanic materials (such as earth or brick dust) are next best. If portland cement can't be avoided, use masonry cement or other mixtures which reduce the overall cement content.

Embodied Energy & Energy Payback

Embodied Energy

Embodied energy is the term used to describe the total amount of energy used in the raw materials and manufacture of a given quantity of a product. The first column on the Product Table is an indication of relative embodied energies. As noted elsewhere, in the absence of information on other aspects of a product's environ-mental impact, embodied energy is often taken to be an indicator of the total environmental impact.

The embodied energy figures below are our best estimates based on a number of sources (1, 18, 21, 22, 23, 24, 25). The difficulty in accurately estimating embodied energy figures is considerable. In our calculations our aim was to preserve the relationship between figures so as to enable comparative judgements, rather than to provide definitive quantities, although we hope too that these are close.

Energy Payback

For a product that insulates or otherwise saves on the energy used in a building, the Energy Payback period of a product is the length of time that it takes for a product to save the same amount of energy as went into its manu-facture (its embodied energy). A shorter payback period indicates a better initial 'investment' of energy.

The energy payback figures presented below cover two scenarios, both calculated using the BREDEM worksheet[20] modelled on a small conventional detached house. In the first, it was assumed that a wall with outer leaf and cavity insulation were already taken into account, which met the current Building Regulations insulation U-value for a house wall of 0.45. The figures show the energy pay-back period in years given an inner leaf of each masonry product. In the second scenario, the wall was assumed to be already 'super-insulated' to U-value = 0.1.

Conclusions

The energy payback figures below show that if a building is to be constructed with close to existing insulation standards, then the choice of masonry material can have a significant impact on the energy use of a building. However, if a building is designed to be 'super-insulating', such insulation levels will never be achieved with the masonry alone - additional insulating layers will be needed. And if sufficient is used to get a U-value of 0.1, the choice of wall material is more sensibly based on choosing the lowest embodied energy.

Inner Leaf Material	Embodied Energy of 100 m2 of wall (GJ)	Energy Payback: original U = 0.45 (in years)	Energy Payback: original U = 0.10 (in years)
Brick	32	51.6	1,067
Dense Concrete Block	20	76.9	2,000
Lightweight Concrete	21	11.7	230
Aerated Concrete	38	11.1	180

	£	Production									Use					ALERT!
	Unit Price Multiplier	Energy Use	Resource Depletion (bio)	Resource Depletion (non-bio)	Global Warming	Ozone Depletion	Toxics	Acid Rain	Photochemical Oxidants	Other	Energy Use	Durability/Maintenance	Recycling/Reuse/Disposal	Health	Other	
Bricks																
Ordinary Clay	1.0	●					•	•	•	•						
Flettons	0.8	•					●	●	•	●						
Soft Mud/Stocks	1.0	●					•	•	•	•						
Perforated Clay	1.0	•					•	•	•	•						
Calcium-Silicate	0.9	●		•	•		•	•	•	●						
Re-Used	1.4															
Concrete Blocks																
Ordinary Dense Blocks	0.3	•	•	●	●		•	•	•	●						
Lightweight Aggregate	?	•	•	•	●		•	•	•	●				●		
Aerated	3.2	●	•	•	•		•	•	•	●						
Composite Insulating	1.4	•	•	●	●	●	•	•	•	●						CFCs?
Stone																
Local	3.2									•						
Imported	?	•								•						
Reclaimed	3.2															
Artificial	1.4	•	•	●	●		•	•	•	●						
Mortar Ingredients																
Ordinary Portland Cement	n/a	●		•	●		•	•	•	●			•	•		Haz. Waste
Pure Lime	n/a	●		•			•	•	•	●				•		
Hydraulic Lime	n/a	●		•	●		•	•	•	●				•		
OP Blastfurnace Cement	n/a	•		•	•		•	•	•	•			•	•		
OP Pulverised Fuel Ash	n/a	•		•	•		•	•	•	•			•	•		
Masonry Cement	n/a	●		•	●		•	•	•	●			•	•		
Sand and Gravel	n/a		•	●						•						

6.4 Product Analysis
6.4.1 Bricks

(a) Ordinary Solid Clay

Production

Energy Use

Brick kilns use a large amount of energy in firing clay bricks, and by their nature a lot of this energy is wasted.[2]

Toxics

Firing bricks often causes toxic gases and vapours to be given off, unless materials are very carefully chosen.[18] Both fluorides and chlorides have specific emission limits set.[26]

Acid Rain

The sulphur content of clays, and therefore the potential for emissions of sulphur dioxide from brick kilns, varies widely, but may be as high as for flettons (see Toxics below).[26]

Photochemical Oxidants

Nitrogen oxides result from the burning of fuel and from other high temperature reactions.[26]

Other

Clays are extracted using conventional mining techniques, which can have a 'profound impact on the local environment'.[2,18]

(b) Flettons

Fletton bricks are made only from Lower Oxford clay and are now made solely by the London Brick Company. They can be considered similar to ordinary clay bricks except for the following:

Production

Energy Use

Fletton clay contains many impurities, which are burnt off in firing, reducing considerably (by up to 75%) the amount of primary fuel used.[4] However, if the fuel content of the clay were included in calculations, the final figure might well be much closer to that of ordinary clay bricks.

Toxics

The impurities in Fletton clay burnt in the kiln result in the potential emission of a wide range of toxic and other pollutants. Of particular concern are mercaptans,[4] fluorides and other halogen compounds, various organic and partial oxidation products, carbon monoxide and particulate matter.[3]

Acid Rain

Sulphur and nitrogen oxides are top of the list of pollutants associated with the Fletton industry.[3,4]

Photochemical Oxidants

Nitrogen oxides result from the burning of fuel and from other high temperature reactions.[3]

Other

A well known association with fletton brickworks is their foul-smelling odour.[3]

Earth Building

Building with earth has been gaining considerable attention recently as a low environmental impact technique. It is an age-old process that is being re-developed for the modern world. There are nearly 50,000 old earth buildings surviving in this country, thus proving the technique's durability and suitability to our climate.[41] Older techniques include wattle and daub, with mud applied to a light wooden frame-work, and cob, where a mud and straw mix is packed by hand to form a thick wall. More modern versions include earth blocks made in a form or press which are then laid like conventional blocks in mud or mortar, and earth rammed into form-work, much like in-situ cast concrete. Small amounts of cement, lime or even bitumen may sometimes be used to stabilise the soil.

Earth building certainly has a low environmental impact compared to conventional building techniques. The earth that is used can often be dug from the site of the building itself, and even when a fairly mechanised approach is used in digging, mixing and placing earth, an earth-built structure has a vastly lower embodied energy than equivalent constructions with brick, concrete or steel.

Protection of the structure from rain-damage is a primary design issue, with large roof overhangs and good detailing being key. Getting building regulations approval for earth built-structures, never mind mortgage finance and insurance, may present some interesting hurdles. Nevertheless there are a number of examples of modern earth building, includ-ing at Hooke Park in Dorset, and at Isle d'Abeau in France where an earth-constructed social housing project was built in 1984.

Building with earth is set to become one of the main ingredients of a truly sustainable architecture of the future, provided that regulations and codes don't hamper progress.

For more information about earth building, see reference numbers (31), (32), (33), (34), (35) and (41). The world centre for information on earth building is CRATerre, BP 53, 38090 Villefontaine, France. Contact Hugo Houbden.

Transport Energy

The energy used per mile per tonne of freight (by road in the UK) averages 0.0056 GJ.[27] This means, for instance, that for a reclaimed brick to have the same embodied energy as a new brick manufactured on your doorstep, it would have to travel an extra 800 miles.

(N.B. This calculation only involves energy use, not the overall environmental impact of road use.)

(c) Soft Mud/Stocks

Bricks made with the soft mud process, often called stocks, and including 'hand-made' bricks, are generally to be considered the same as ordinary clay bricks, except for the following:

Production

Energy Use

The clay used in the soft mud process often requires additional drying by heated air before firing.[1]

(d) Perforated Clay

Perforated clay bricks, otherwise similar to ordinary solid clay bricks, have perforations running through them from top to bottom. This means reduced resource usage per brick, reduced processing time and energy, and reduced weight (and therefore transport energy). Perforated clay bricks may also have a slightly higher insulation value than solid bricks.[1,30]

(e) Calcium-Silicate

Calcium-silicate, sand-lime or flint-lime bricks are made with suitable aggregates plus lime.[1] The impacts of both these are combined here, plus:

Production

Energy Use

Calcium-silicate or sand-lime bricks have to be 'autoclaved' - cooked in steam at high pressure.[1]

Resource Depletion (non-biological)

Sand suitable for the manufacture of calcium-silicate bricks ''is not so plentiful in the UK''.[1]

(f) Reclaimed Bricks

Reclaimed bricks are in many ways the perfect environmental option, and there is a thriving market in bricks reclaimed from demolished Victorian buildings. There is a strong argument that it is environmentally better to re-use and repair buildings rather than demolish them and just re-use the salvageable parts. But we have come across no evidence that economic demand for reclaimed bricks is actually encouraging the demolition of buildings that would otherwise be repaired.

6.4.2 Concrete Blocks

(a) Ordinary Dense

The production of ordinary dense concrete blocks requires cement plus sand and aggregates. The impacts of these (see below for details) have been combined here, plus:

Production

Energy Use

Most concrete blocks are steam cured, sometimes under pressure.[1]

(b) Lightweight Aggregate

These share the same impact as for ordinary dense concrete blocks, except for:

Production

Energy Use

The manufacture of most lightweight aggregates (expanded clays and shales, sintered pfa, exfoliated vermiculite and expanded perlite) involves high energy input as the raw materials are sintered or fired in kilns. The exceptions are foamed or pelletised slag, which is treated whilst still hot from the blast-furnace, pulverised-fuel ash (un-sintered) and furnace clinker (breeze).[1]

The lighter weight of the blocks does to some extent reduce the energy expenditure related to transport.

Other

The process used to convert vermiculite and perlite into light-weight aggregates can cause problems with dust and particulate emissions.[28]

Health Hazards

According to Christopher Day (quoting Swedish research), blocks made with pumice, blast-furnace slag or pulverized-fuel ash may cause a radiation hazard, which could be responsible for an increased risk of cancer, although the UK Quality ash Association points out that "the use of ash in building materials results in negligible increase of radiation dose from living in dwellings".[48] Day suggests that the production of these materials concentrates the natural uranium present in the deep earth rock, but this will clearly vary according to the source of the materials.

When used in building blocks to build a whole house, average radiation levels have been measured to be equal to the 'action level' for radiation from naturally-occurring radon gas specified by the Building Regulations (200 Bq/m^3). Exceptional measurements have reached four times this level.[36]

It would be misleading to suggest that fly ash alone is responsible for radiation in concrete blocks as other materials which fly ash replaces could also have a similar effect. The emission and diffusion of radon is highly variable and can be affected by the design and use of other materials.[49] There are also widely varying opinions as to what can be regarded as safe levels of radiation and these must be offset against the environmental benefits of using by-product materials such as fly ash.

(c) Aerated

Aerated concrete blocks are made from a cement/lime/sand slurry, with a small amount of aluminium sulphate powder added. The aluminium reacts with the lime, forming hydrogen bubbles, and the block rises like a cake. The block is cut and then 'autoclaved' in steam at pressure to develop strength.[1] The impacts of cement, lime and sand have been combined here. Sometimes pulverised-fuel ash is used instead of sand, and this would reduce the impact accordingly, but perhaps raise the same radiation issue as for lightweight aggregate blocks (see facing page).

Production

Energy Use

Though only used in small quantities here, aluminium is a very high embodied energy material.[18]

Global Warming

Ordinarily, concrete does not carbonate significantly, but aerated concrete with its open texture does absorb a significant amount of CO_2 from the atmosphere. This is sometimes forced during manufacture with waste gas from other industrial processes.[13] This reduces the overall global warming impact.

(d) Composite Insulating Blocks

Production

Concrete blocks made with attached insulation inherit the same criticisms as the concrete block they are based on, plus:

Ozone Depletion

According to the AECB, there are no blocks available with attached insulation that are made without ozone-depleting chemicals such as CFCs.[2] However, the production of CFCs after 1994 has been outlawed by the European Union. It is not yet clear which processes have been or will be substituted.

6.4.3 Stone

(a) Local

Production

Energy Use

Energy use is limited to the fossil fuels used in quarrying, shaping and transporting stone.[43] Stone from a local source will therefore have the lowest energy cost. Because of high wastage rates the energy requirements for finished stone products can be as high as 4MJ/kg of stone produced, but may be considerably less depending on extraction methods.[42]

Resource Use (non-bio)

Because of the requirement to produce stone of given dimensions and quality, reject material can form a large percentage of quarry production.[42]

Resource Use (bio)

Quarrying has the potential to cause significant landscape and ecological damage unless carried out with extreme care.

Whilst stone can be extracted with minimal impact, it is often criti-cised, especially the extraction of limestone in National Parks.[18]

Global Warming

It is estimated that CO_2 emissions from production plant, quarry transport and electricity are of the order of 0.53 tonnes CO_2 per tonne of finished product.[42] This figure does not include transport to the point of use.

Other

Quarrying can cause local impacts such as noise, dust and vibration, plus increased heavy road traffic.[42]

Use

Durability

Stone is highly durable.

Recycling/Disposal

The extremely high durability of stone allows active reclamation and re-use.[42]

(b) Imported

Stone that has to come a long way to the end user has the same impact of local stone, plus:

Production

Energy Use

Stone is only a relatively low energy product if it doesn't have to travel far. Most stone is more dense than brick or concrete blocks, and long distance transport is very energy intensive.[18]

(c) Reclaimed

The same arguments apply here in favour of reclaimed stone as for reclaimed bricks above.

(d) Artificial

Artificial stone uses stone waste as a facing to concrete blocks.[18] It therefore has practically the same impact as ordinary concrete blocks (see facing page).

Ready Mix?

According to the Association for Environment Conscious Builders (the AECB), ready-mixed mortars and concrete may present least risk to the environment. The mixing and storing of cements on building sites is prone to accident, spillage and wastage, whilst ready-mixed products can be made under much more controlled conditions.[2]

6.4.4 Mortar

(a) Ordinary Portland Cement

Production

Energy Use

Cement kilns must be fired at very high temperatures (~1400°C) using fossil fuels.[1,2,18] The embodied energy of OPC is 6.1MJ kg[-1] (wet kiln production) or 3.4Mj kg[-1] (dry kiln production).[42]

Transport costs of cement are high due to its volume and density. Sadly, more than 75% of cement production leaves the plant by road. The works at Dunbar, Hope and Weardale (all owned by Blue Circle) deserve commendation for using rail transport for 50% or more of output (1989 figures).[7]

Resource Depletion (non-bio)

The main raw materials for cement - limestone or chalk - are fairly widespread in the UK. However, in some areas, notably the south-cast, suitable 'permitted' reserves are running low.[7]

Global Warming

The manufacture of cement from chalk or limestone involves a chemical reaction in which carbon dioxide is given off at a rate of 500kg tonne[-1]. The cement industry is the only significant CO_2 polluter other than fossil fuel burning - responsible for about 450 million tonnes, or about 8-10% of the global total.[5]

Toxics

OPC contains heavy metals, ''of which a high proportion are lost to the atmosphere'' on firing.[2] Organic hydrocarbons and carbon monoxide are also released, and fluorine compounds can also be present.[6]

Acid Rain

Sulphur dioxide is produced in the cement kiln, both as part of the chemical reaction of the raw materials, and as a product of burning fossil fuel. Normally, however, it is mostly re-absorbed into the cement by chemical combination, and only a small amount escapes. Nitrogen oxides from fuel burning are not absorbed.[6]

Photochemical Oxidants

Nitrogen oxides result from the burning of fuel and from other high temperature reactions.[6]

Other

The production processes can cause a serious dust problem which is hard to control.[2] Again, the extraction of limestone is cause for concern locally.[18]

Admixtures, added in small quantities to concrete or mortar in order to alter its workability, setting strength and durability. These are discussed in detail in chapter 12, page 150.

Use

Recyclability

The higher strength of mortars based on ordinary portland cements often leads to mortars being used that are considerably stronger than are needed for structural reasons. This means a lot harder work when cleaning up bricks for re-use, if indeed they can be reclaimed at all.[29]

Health Hazards

On the building site, OPC dust may contain free silicon dioxide crystals (the cause of silicosis), the trace element chromate (a cause of stomach cancer and skin allergies), and the lime content may cause skin burns.[2] The fast setting/frost proofing additive calcium chloride which is sometimes used can cause skin ulceration or burns.[2]

Alert

Hazardous waste burning

All the major manufacturers are experimenting using substitute 'fuels' derived from hazardous or toxic chemical wastes. There is a strong economic incentive to do so, as

---

Lime Mortars

In addition to the benefits of lime over portland cement in the environmental impact of production, lime mortars have other advantages. Lime mortars are softer and less brittle than cement mortars - and so are better suited to materials such as softer stone, bricks and timber, which might otherwise suffer cracking. Lime mortars are also more porous - allowing a wall to 'breathe' and to dry out when wet. Lime mortars have been used for centuries, in many buildings that have lasted for centuries, so durability is not a problem. Yet if and when demolition occurs, the softer lime mortar is much easier to remove when reclaiming bricks or stone.

Disadvantages of lime mortars over cement mortars include the slow curing time, the length of time before frost resistance is achieved, and possibly cost as lime putty, though cheap, is not widely available.

The preparation and use of lime mortar is sufficiently different to that of portland cement mortars to merit some study - but lime mortars have been used for 2,000 years, and no particularly difficult 'new' skills have to be learnt by the builder. The key difference for the bricklayer or mason is that lime mortar takes much longer than cement mortar to set. Time must therefore be allowed for each layer to set sufficiently before the next is added. The mixing and preparation of lime mortars is also different to, but no more difficult a process than for cement mortars.

For detailed information, see references no 38, 39 and 40, available from the Society for the Protection of Ancient Buildings (SPAB), 37 Spital Square, London E1 6DY (tel: 0171 377 1644).

Thornton Kay from SALVO explains why we should

Reclaim Old Building Materials

There is a protocol for recycling that has been accepted by both the UK Government and the European Parliament which can be applied to building materials in the following way:

1. Re-use a building without demolition, but if demolition in whole or in part is to be carried out then:
2. Reclaim whole components such as doors and bricks. If this is not possible then:
3. Re-cycle and re-manufacture components such as bricks and masonry into hardcore, or plastics into a new product, such as motorway cones. Failing which in the last resort:
4. Beneficially destroy, for instance by burning wood for energy recovery in a power station.

Around 10 or so old bricks embody the energy equivalent of a gallon of petrol, and while 3.5 billion bricks are manufactured annually in the UK, 2.5 billion are destroyed. If bricks are crushed and used as hardcore, then they are considered to have been recycled, but all the energy is lost. If bricks are re-claimed whole to be re-used as bricks, then the energy content is preserved. This example shows that reclamation is better for the environment than recycling.

In Britain around 25,000 tonnes of reclaimable materials are disposed of every day into landfill sites. Around half of this material is from Edwardian and Victorian buildings, and represents a period when materials were made to last. For over a hundred years the UK was the biggest importer of the world's finest timber, high quality bricks were manufactured, quarrying skills were second to none, and metal and glass manufacture was at its peak.

If you buy a new BMW car around 70% of it is made from recycled material, but if you buy a new building usually less than 1% is reclaimed.

SALVO suggests that:

● specifiers should use at least 50% of reclaimed material in every building and landscaping project
● local reclamation dealers should be used, and
● reclaimable materials should never be landfilled.

SALVO is a two person information service for reclaimed building materials, run by Thornton Kay and Hazel Maltravers. Our publications are:

Salvo Monthly [sub £20 for 10 issues]

Water Bylaws Factsheet [£1.50 inc. p&p]

Reclaimed Roof Tiles Poster [£2.50 inc. p&p]

The Salvo Pack [£5.75 inc. p&p] includes:

24 page listing of 300 UK dealers

Reclaimed Roof Tile Poster

County Dealer Listings for 3 counties of your choice

Free copy of Salvo Monthly

Additional County Dealer Listings [£1.00 inc. p&p]

Contact: SALVO, Ford Woodhouse, Berwick upon Tweed TD15 2QF (Tel: 01668 216494) or email: tkSALVO@delphi.com

---

they are paid to take these materials off the hands of their producers, as well as saving in coal or gas, and thus the cash savings are considerable. Cement kilns operate at higher temperatures than most specialised toxic waste incinerators, but this 'recycling' of wastes still causes concern because there are much less stringent controls on cement kilns. Although hazardous waste burning is still only undergoing 'trials' at the moment, environmentalists are concerned that the industry itself seems to be the judge and jury of these trials. The main concern is with toxic emissions to the air, of dioxins and heavy metals. It is also possible that many of the chemicals are absorbed into the cement itself,[6] but there has been no published investigation into this. No manufacturers label such products for the consumer to identify. The campaigning group Communities Against Toxics is focusing on this issue.

(b) Pure Lime

In this study, pure lime includes any non-hydraulic lime, and lime without any pozzolanic additives. (See hydraulic lime, p.60) It may be in the form of quicklime, lime putty or powdered 'bag lime'. Most lime produced in the UK is non-hydraulic.[39] See Box on 'Lime Mortars' for more information.

Production

Energy Use

Lime kilns must be fired at high temperatures, from 900°C up to 1100°C,[8] but not as high as that required for cement firing.

Resource Depletion (non-bio)

The main raw materials for lime, limestone or chalk, are fairly widespread in the UK. However, in some areas, notably the south-east, suitable 'permitted' reserves are running low.[7]

Global Warming

The manufacture of lime from chalk or limestone involves a chemical reaction in which carbon dioxide is given off in large quantities, just as with cement. However, to some extent lime re-absorbs this carbon-dioxide as it sets. A pure lime mortar in not too large a section will re-absorb most of its CO_2. Deep sections may never set entirely, due to the atmospheric gases being unable to reach right inside, but they will continue to re-absorb CO_2 slowly throughout their life.[1,8,9,51]

Toxics

Carbon monoxide and fluorine compounds can be present in emissions.[10]

Acid Rain

Similar comments apply to lime production as to ordinary portland cement.[10]

Other

Dust and particulate matter are common emissions from lime plant.[10] The extraction of lime-stone also has considerable impact on the local environment.[18]

Use

Health Hazards

Lime may cause skin burns to building workers when wet.[2] If quicklime is slaked (added to water) on site, the process is potentially dangerous both because of the heat and splashing of caustic materials, and because of the exposure to lime dust in the steam given off. Industrial lime slaking comes under local authority air-pollution control because of this.

(c) Hydraulic Lime

Included here are limes that are either made from limestone or chalk with natural clay impurities, or pure lime with 'pozzolanic' additives blended at a later stage. Hydraulic limes are so called because they can set under water (and should not be confused with hydrated lime). Hydraulic lime shares the same environmental rating as pure lime with the following exception:

Production

Global Warming

Whilst responsible for a similar amount of CO_2 in production, hydraulic lime mortars achieve their set by more complex chemical reactions than pure lime mortars, and don't re-absorb the same amount of CO_2 on setting.

(d) OP Blastfurnace Cement

Blastfurnace slag, a waste product of the iron smelting industry, can be added to ordinary portland cement. In Japan, cement with 50% blastfurnace slag is awarded the Japanese Ecomark.[15] The use of this product reduces the overall amount of cement needed, and hence reduces the overall impact of the product.

Health Hazards

Blastfurnace slag may contain some residual radiation-forming material - see Lightweight Blocks above - but the amounts used just in the cement for a building are small compared to the amount of aggregate in block-built walls, so the risk is presumably much less.

(e) OP Pulverised-Fuel Ash Cement

The ash waste from pulverised-fuel (coal) power stations is sometimes used as a cement extending/additive material. The use of this product also reduces the overall amount of cement needed, and hence reduces the overall impact of the product.

Health Hazards

Pulverised-fuel ash may contain some residual radiation-forming material - see Lightweight Blocks above - but the amounts used just in the cement for a building are small compared to the amount of aggregate in block-built walls, so the risk is presumably much less.

(f) Masonry Cement

Masonry cement is made with a minimum of 75% ordinary portland cement plus an inert filler, usually fine ground limestone.[9] This will reduce the overall impact of the cement, as less processing is required.

(g) Sand & Aggregates

Production

Resource Depletion (bio)

Dredging of aggregates may lead to coastal erosion and loss of habitats for fish or other marine life.[16]

Resource Depletion (non-bio)

Sand and aggregates mainly come from land quarries and marine dredging. However, as gravel reserves deplete, alternatives such as crushed rock are becoming more widely used.[1,10] It is getting ever harder to find new quarry sites that would not result in unacceptable environmental damage.[18] 'Superquarries' have come in for particular criticism.[17]

Other

The mining or dredging of sand and aggregates can have significant local impact, in terms of landscape change, noise, dust, traffic etc.[11]

Restoring Quarries?

Although many worked-out quarry sites are being restored as part of the industry's environmental duty, some of the future uses of sites may not always meet with environmentalists' wholehearted approval. Take for instance the 2,500 acres of land used to quarry chalk by Blue Circle cement at Ebbsfleet in Kent. This site will be 'restored' with a new town, a channel tunnel high-speed line station, and Europe's biggest shopping centre! Other sites are often used as waste tips.

Waste Incineration in Cement Kilns.

Concern is growing over the possible toxic effects of burning Secondary Liquid Fuels in cement kilns,[46] also known as Waste Derived Fuel or Recycled Liquid Fuel.[46]

The main incentive for using these fuels is a reduction in fuel costs, which account for some 35% of the cement industries direct manufacturing costs. Cement companies burning SLF currently receive up £15 per tonne from waste producers, who view this as a cheap disposal option, giving the cement companies a strong economic incentive to burn these fuels.[46]

The Issues:

It is argued that the use of SLF as a fuel may divert solvents from materials recovery, which is environmentally preferable to burning for heat recovery. This is denied by the industry, who claim that the majority of SLF burned in cement kilns is solvents with a low recycling capacity, which are diverted from landfill rather than from materials recovery. Energy recovery is generally considered a less preferable environmental option to materials recovery, but preferable to landfill.[46]

It is uncertain whether emissions from burning SLF are significantly different to those from burning coal, and it is reported that HMIP monitoring data is currently of insufficient quality to draw any firm conclusions regarding emissions.[46] The composition of waste derived fuels can be highly variable, often containing PCBs, organohalogen compounds and heavy metals, and there are concerns over the uncontrolled emission of heavy metals and other toxics. The industry argue that the high temperatures (higher than most specialist toxic waste incinerators), long residence times and alkali conditions in cement kilns facilitate virtually complete combustion of toxic organics and immobilisation of heavy metals.[46,44] However, Friends of the Earth argue that SLF burning kilns should be subject to the same emission controls as hazardous waste incinerators[46] due to concerns about dioxin and heavy metal emissions, and the high variation in stack emissions from burning SLF. They point to variation in temperature, turbulence and insufficient oxygen in cement kilns, as potential barriers to complete combustion of the waste.[48] A major concern of the environmentalists is that the industry itself seems to be the judge and jury of these waste burning trials.

If it is classed as a waste, as has been recommended by the DoE, SLF will be subject to the same "duty of care" regulations as hazardous waste. More importantly, cement kilns burning SLF would become subject to stringent new EC emissions standards. They would also be exposed to the planning process, making them more locally accountable.[46]

Health Effects of Burning SLF

There are no confirmed links with cement manufacture and health problems possibly due to a lack of research in this area. Fears have been raised by recent findings, such as a high incidence of asthma in residents living downwind of a Blue Circle cement works in County Durham, which has recently started trial burns of industrial waste.[45] Residents around the Castle Cement works in Clitheroe have complained of headaches, hallucinations, skin rashes, sore throat and prickly eyes, since the cement works started burning Waste Derived Fuel in mid- 1992. With "soaring asthma rates" amongst children, 11 families in Clitheroe won legal aid in June 1995 to pay for air quality monitoring and to claim compensation from Castle Cement.[47] It is also possible that many of the chemicals are absorbed into the cement itself but there has been no published investigation into this.

Users of Waste Derived Fuels[44]

Company	Location	Fuel
Caste Cement	Ribblesdale	Cemfuel
	Ketton	
Cemfuel		
	Padeswood	Cemfuel
Rugby Cement	Barrington	SLF
	Southam	SLF
Blue Circle	Cauldon	Tyre Chips
	Weardale	SLF
	Dunbar	SLF
	Claydon	SLF
Redland	Thrislington	solvent
	Whitwell	SLF

6.5 Specialist Suppliers

6.5.1 Reclaimed Bricks

Arborfield Services, Woodlands Farm, Woodlane, Barkham RG41 4TN. (tel: 01734 760244)

Au Temps Perdu, 5 Stapleton Road, Easton, Bristol BS5 0QR (tel: 01179 555223)

Cawarden Brick Company, Cawarden Springs Farm, Blithbury, Rugeley, Staffs. WS15 3HL (tel: 01889 574066)

Cheshire Brick & Slate Co, Brook House Farm, Salters Bridge, Tarvin Sands, Chester CH3 8HL (tel: 01829 740883 fax: 01829 740481)

Conservation Building Products Ltd, Forge Works, Forge Lane, Cradley Heath, Warley, W. Mids. B64 5AL (tel: 01384 564219)

Dorset Reclamation, The Reclamation Yard, Cow Drove, Bere Regis BH20 7JZ (tel: 01929 472200 fax: 01929 472292)

Romsey Reclamation, Station Approach, Romsey Railway Station, Romsey, Hants. SO51 8DU (tel: 01794 524174 Fax: 01794 514344)

Ronsons Reclamation & Restoration, Norton Barn, Wainlodes Lane, Norton, Gloucester GL2 9LN (tel: 01452 731236 Fax: 01452 731888)

Solopark Ltd, The Old Railway Station, Station Road, Nr Pampisford, Cambs. CB2 4HB (tel: 01223 834663 Fax: 01223 834663)

Symonds Bros., Winton Oast, Puddingcake Lane, Rolvenden, Cranbrook, Kent TN17 4JS (tel: 01580 241313)

John Walsh & Sons, Lyntown Trading Estate, Old Wellington Road, Eccles, Manchester M30 9QG (tel: 0161 789 8223 fax: 0161 787 7015)

6.5.2 Reclaimed Stone

Barnes Building Supplies, Moor Lane Trading Estate, Sherburn-in-Elmet, Leeds LS25 6ES (tel: 01977 683734)

Cardiff Reclamation, Site 7, trenorta Ind. Estate, Roverway, Cardiff CF2 2SD (tel: 01222 458995)

Clayax Yorkstone Ltd, Derry Hill, Menston, Ilkley, W. Yorks. LS29 6AZ (tel: 01943 878351 fax: 01943 870801)

Gallop & Rivers, Ty'r'ash, Brecon Road, Crickhowell, Powys NP8 1SF (tel: 01873 811804)

J A T Environmental Reclamation, The Barn, Lower Littleton Farm, Winford Road, Chew Magna, Avon (tel: 01275 333589)

Reclamation Services Ltd, Catbrain Quarry, Painswick Beacon, Painswick, Glos., GL6 6SU (tel: 01452 813634)

Searles Ltd, The Yard, Trenders Ave, Rayleigh, Essex SS6 9RG (tel: 01268 780150)

Stone Brokers, Greenbanks, Dalkey Avenue, Dalkey, Co Dublin, Eire

Walcot Reclamation Ltd, Unit 8A, The Depot, Riverside Business Park, Lower Bristol Road, Bath BA2 3BD (tel: 01225 484315)

There are a very large number of suppliers of reclaimed building materials, too many to list them all here. The above is only a selection of some major ones. For a pack of information including a full listing of suppliers in your area and a copy of SALVO News send £5.75 to SALVO, 1 The Cottage, Fordwoodhouse, Berwick upon Tweed TD15 2QF tel: 01668-6494.

6.5.3 Quick Lime & Lime Putty

ARC Southern, Battscombe Quarry, Cheddar, Somerset BS27 3LR (tel: 01934 742733 fax: 01934 742956) *(quicklime)*

R H Bennett Lime Centre, Near Winchester, Hants. (tel: 01962 713636) *(quicklime & putty)*

Rose of Jericho at St Blaise Ltd, the Works, Westhill Barn, Evershot, Dorchester, Dorset DT2 0LD (tel: 01935 83662/3 fax:01935 83017) *(quicklime & putty)*

Bleaklow Industries Ltd., Hassop Ave, Hassop, Derbyshire DE45 1NS (tel: 01246 582284 fax: 01246 583192) (quicklime & putty)

H J Chard & Sons, Albert Rd, Bristol, Avon BS2 (tel: 01179 777681 fax: 01179 719802) *(quicklime & putty)*

Tilcon (South) Ltd, Tunstead Quarry, Wormhill, Buxton, Derbyshire SK17 8TG (tel: 01298 768444 fax: 01298 768334) *(quicklime)*

Masons Mortar Ltd., 61-67 Trafalgar Lane, Edinburgh EH6 4DQ (tel: 0131 555 0503 fax: 0131 553 7158) *(putty)*

Potmolen, 27 Woodcock Ind. Est., Warminster, Wiltshire BA12 9DX (tel: 01985 213960 fax: 01985 213931) *(putty)*

RMC Indusrial Minerals Ltd (Peakstone Lime), Hindlow works, nr. Buxton, Derbyshire SK17 0EL (tel: 01298 72385) *(quicklime)*

Severn Valley Stone, Tewksbury, Glos., (tel: 01684 297060) *(putty)*

Singleton Birch Ltd, Melton Ross Quarry, Barnetby, north Lincolnshire DN38 6AE (tel: 01652 688386) *(quicklime)*

Tamar Trading Co Ltd., 15 Bodmin Street, Holdsworthy, Devon EX22 6BB (tel: 01409 253556) *(putty)*

Tilcon Ltd Central Ordering Service, Sevenoaks Quarry, Bat and Ball Road, Sevenoaks, Kent TN14 5BP (tel: 01732 453633 fax: 01732 456737) *(putty)*

Whitford Sand Lime and Mortar Co, Bedw Cottage, Whitford Road, Holywell, Clwyd, CH8 9AE (tel: 01352 714144) *(putty)*

Rory Young, 7 Park Street, Cirencester, Glos GL7 2BX (tel: 01285 658826) *(putty)*

6.6 References

1. Construction Materials Reference Book (ed. D K Doran) Butterworth-Heinemann Ltd, Oxford, 1992

2. Greener Building Products & Services Directory (K Hall & P Warm) Association for Environment Conscious Building, Coaley

3. Chief Inspector's Guidance to Inspectors, Process Guidance Note IPR/3/6 - Ceramic Processes (Her Majesty's Inspectorate of Pollution) HMSO, London, 1992

4. ENDS Report no. 227 December 1993

5. The World Environment 1972-1992 - Two Decades of Challenge (eds. M K Tolba et al) UNEP/Chapman & Hall Ltd, London, 1992

6. Technical Note on Best Available Technologies Not Entailing Excessive Cost for the Manufacture of Cement (EUR 13005 EN), Commission of the European Communities, Brussels, 1990

7. Minerals Planning Guidance: provision of raw material for the cement industry (Department of the Environment & Welsh Office) 1991

8. Production and Use of Lime in the Developing Countries (Overseas Building Note 161), G E Beasey (Building Research Establishment, Garston) 1975

9. Bricks and Mortar (Overseas Building Note No. 173), Building Research Establishment, Garston, 1977

10. Chief Inspector's Guidance to Inspectors, Process Guidance Note IPR/3/2 - Lime Manufacture and Associated Processes (Her Majesty's Inspectorate of Pollution) HMSO, London, 1992

11. Minerals Planning Guidance: Guidelines for Aggregates Provision in England (Department of the Environment) 1994

12. Alternatives to OPC, Overseas Building Note 198 (R G Smith, Building Research Establishment, Garston) 1993

13. Lightweight Concrete, A Short & W Kinniburgh

14. Northern Echo 18th July 1994

15. Ecolabelling of Building Materials and Building Products; BRE Information Paper IP 11/93 (C J Atkinson, M A & R N Butlin, Building Research Establishment, Garston) May 1993

16. Poisoners of the Sea (K A Gourlay) Zed Books, London 1988

17. New Scientist 8/1/94

18. Eco-Renovation - the ecological home-improvement guide (E Harland, Green Books Ltd, Dartington) 1993

19. BREEAM/New Homes Version 3/91 - An Environmental Assessment for New Homes (J Prior, G Raw & J Charlesworth, Building Research Establishment, Garston) 1991

20. Energy Assessment for Dwellings using BREDEM Worksheets; IP 13/88 (B R Anderson, Building Research Establishment, Garston) 1988

21. Manufacture and Application of Lightweight Concrete (Overseas Building Note No. 152) (Building Research Establishment, Garston) 1973

22. Design with Energy - the Conservation and Use of Energy in Buildings (J Littler & R Thomas, Cambridge University Press, Cambridge 1984)

23. Building Energy Code (CIBSE) 1982

24. Energy Efficient Building (Susan Roaf & Mary Hancock, Blackwell Scientific Publications, Oxford) 1992

25. Environmental Building (Pat Borer, Centre for Alternative Technology, Machynlleth) 1994

26. Secretary of State's Guidance - Manufacture of heavy clay goods and refractory goods (EPA 1990 part 1) (HMSO, London, 1991)

27. Calculated from 1992 figures in: Transport Statistics Great Britain 1993 edition, DoT, Govt Statistics Service

28. Secretary of State's Guidance - Exfoliation of vermiculite and expansion of perlite (EPA 1990 part 1) (HMSO, London, 1991)

29. Discussion with Peter Horsley of Conservation Building Products Ltd. (Cradley Heath) October 1994

30. Perforated Clay Bricks, Digest 273 (Building Research Establishment, Garston) 1983

31. Stabilised Soil Blocks for Building, Overseas Building Note 184 (M G Lunt, Building Research Establishment, Garston) 1980

32. Bricks and Blocks for Low Cost Housing, Overseas Building Note 197 (R F Carroll, Building Research Establishment, Garston) 1992

33. Appropriate Building Materials - A Catalogue of Potential Solutions (Third Revised Edition) (R Stulz, K Mukerji, SKAT Publications, Switzerland) 1993

34. Building For a Future (AECB, Coaley) Summer 1992

35. Building For a Future (AECB, Coaley) Summer 1994

36. Places of the Soul: Architecture and Environmental Design as a Healing Art (Christopher Day, The Aquarian Press (Thorsons), Wellingborough) 1990 (quoting research from Radon i bostader, Socialstyrelsen, statens planverk, statens stralskyddinstitut, Sweden, 1988)

37. Spon's Architects' and Builders' Price Book 1995 (E & F N Spon, London) 1994

38. Lime in Building: a Practical Guide (Jane Schofield, Black Dog Press, Devon) 1994

39. An Introduction to Building Limes (Michael Wingate, Society for the Protection of Ancient Buildings: Information Sheet 9)

40. Using Lime (Bruce & Liz Induni, Lydeard St Laurence) 1990

41. Earth building today (Sumita Sinha & Patrik Shumann in The Architects' Journal, 22 September 1994)

42. The Environmental Impact of Building and Construction Materials Volume B: Mineral Products. (R. Clough & R. Martin). Construction Industry Research and Information Association, London. June 1995.

43. Environmental Building News 4 (4). July/August 1995

44. Hazardous Waste and Cement Kilns. Friends of the Earth Briefing Sheet, June 1995.

45. ENDS Report 244 p.22-25 May 1995

46. ENDS Report 245, p.27-29 June 1995

47. Incineration by the Back Door - Cement Kilns as Waste Sinks. (T. Hellberg). The Ecologist, 25 (6) p.232-237. November/December 1995

48. Radioactivity of combustion residues from coal fired power stations. (Puch K.H. Keller G & vom Berg W.). VGB PowerTech Vol. 8 p623-629. 1996

49. Coal Ash Reference Report. Thermal Generation Study Committee - Group of Experts: Thermal Power Plant Residue Management. UNIPEDE, Paris. 1997

Timber 7

7.1 Scope of this Chapter

This chapter looks at the environmental impact and sustainability of timber. Its aim is not to compare timber to other materials with similar functions, but rather to assist in choosing the (environmentally) best timber for the job, and therefore we have not included the usual Product Table in this chapter.

7.2 Introduction

Sustainability is at the top of the agenda for timber, as for no other industry. Perhaps because it is such a natural product, 'good wood' has been the focus for both campaigners' demands and industry's claims for many years. And due to increased awareness of the damage to rainforests caused by logging, there has been a decline in tropical timber consumption in the UK in recent years.[13] The majority of the world's timber producers may claim that their production is sustainable, but as we will see, sustainability has a wide range of meaning. Only a small proportion of the industry can justifiably be described as being anywhere near to 'truly sustainable' in its widest sense.

The quantity of wood consumed is also an issue. Just because it grows on trees doesn't mean that we can use it without restraint. There will always be an environmental impact, even with sustainably produced timber, and the usual green rules of 'buy locally' and 'reduce, re-use, repair and recycle' still apply.

It is heartening, though, that we *can* buy timber produced from well-managed forests, and which is verified as such by credible independent expert bodies such as the Soil Association. If some other industries were to make even this small progress, we might really be getting somewhere.

However, the credible certification of timber is only just beginning to take off, and specifiers unfortunately may still have to do a bit of work asking questions of suppliers in order to ensure they are getting the best wood they can.

Greenie Points

BREEAM/New Homes version 3/91 awards 1 credit for specifying timber and timber products for use as an integral part of the building (e.g. structural wood, window frames, architraves) which are **either** *from well managed, regulated sources* **or** *of suitable reused timber. (Plus 1 credit for non-integral parts similarly sourced, e.g. wardrobes and kitchens).*

"The environmental benefits of timber are overstated [by the timber industry's Forests Forever and Think Wood campaigns].. all the evidence gathered over many years by various international agencies [shows] that locally, regionally and globally, the timber industry is operating on a non-sustainable basis."
Friends of the Earth [15]

Sustainability means different things to different people. For much of the forestry industry, a claim to be operating sustainably may only be referring to economic sustainability or 'sustained yield'. This is concerned with ensuring the extraction of a continuing financial return or monetary value of timber.

Whilst of course important, economic sustainability does not satisfy the demands of all. It does not consider the preservation of the biological resources of the original forest system, the maintenance of environmental services such as watershed management and soil conservation, recognition of indigenous peoples' rights, or the provision of jobs, food and materials for the local community. These might come under terms such as biological-, environmental- and/or social-sustainability.[29]

Truly sustainable forestry in the widest sense is probably an unattainable ideal, since any activity in a forest will have effects such as reducing the numbers of some species (although other species may increase). A study carried out for the International Tropical Timber Organisation showed that, using the few case studies with sufficient information, less than 1 per cent of tropical moist forest was under sustained-yield management for timber production.[35]

Attention has often been more on tropical forestry and logging practices, and we have found no comparable figures for elsewhere in the world, but it is likely that plantation forestry may also not be sustainable if it requires continuing inputs of fertilisers and pesticides to achieve sustained yields. According to the professor of forest ecology at the University of British Colombia, Hamish Kimmins: "Nowhere in the world have we had three rotations of forests without the ecosystem collapsing".[19] Nevertheless, because of the long rotation lengths in forestry, there is as yet little information to predict what the long-term outcomes of different management practices will be. Because of this, rather than trying to define sustainability, campaigners' efforts are now focusing on developing guidelines and criteria for responsible forestry practices based on a precautionary approach, with the aim of minimising the environmental and social impacts of timber production as far as can be judged on present information.

"All materials have individual qualities. Wood is warm, it has a life to it even though the tree is long felled.... It is hard to make a cold-feeling room out of unpainted wood, hard to make a warm soft approachable room out of concrete."
Christopher Day [16]

WOODMARK
THE SOIL ASSOCIATION'S CERTIFICATE OF RESPONSIBLE FORESTRY

Timber Labels - A Pair of Woodmarks

Somewhat confusingly, there are two eco-labels known as the Woodmark. The first is awarded by the Soil Association, who are well known for their organic food certification scheme and is accredited by the Forest Stewardship Council (FSC). (See page 76.) For a list of timber suppliers who have been awarded this Woodmark, see Certified Sources on page 78.

The other Woodmark comes from the Forestry Industry Committee of Great Britain (FICGB). They say it is used by thousands of timber producers and suppliers, but it simply indicates that timber has come from UK source, and that the source complies with the standard Forestry Authority rules. It is a 'rule-based' system that suppliers apply themselves. No site visits or inspections are performed.

Tree felling in the UK is controlled by the Forestry Act 1967 and the Forestry (Exceptions from Restrictions of Felling) Regulations 1979. In most cases it is an offence to fell trees without an agreed plan of operations or a felling licence, most of which carry a requirement to restock with suitable species.[7]

The British Timber Merchants Association claims that there is 'little difference in standards' between the FICGB Woodmark and the Soil Association one.[28] But the Soil Association's Woodmark is awarded after inspections and site-visits, while the FICGB version works as self-declaration by suppliers. The SA's (expected) accreditation by an internationally recognised independent body, and the international applicability of its Woodmark, make its scheme more likely to be trusted by consumers and environmentalists.

The following section looks at the environmental impact of timber production generally. Where possible, distinctions are drawn between different types of timber and regions of production, but we have not found enough information in order to perform a separate analysis for each type of timber or region. The Buying Timber section attempts to draw some conclusions from this analysis.

7.2.1 Transport Energy

Whilst the main energy input in tree production may be from sunlight via photosynthesis, transport energy, supplied by fuel oil, is the most important energy cost for timber in terms of environmental impact. Timber is very much a world-wide traded commodity, and some literally does come to the UK from the other side of the globe. The table shows an approximation of the energy used in fuel oil to transport timber to the UK from various parts of the world. Whilst container ships are a relatively energy efficient means of bulk transport, the vast distances involved mean that the 'embodied energy' of imported timber can add up to a significant amount. (And no-one has yet started importing timber with sailing ships!)

As a guide to understanding these quantities, we have also listed the embodied energy figures of some other common building materials. But remember that this is not a direct comparison of like for like - one tonne of timber is not usually a substitute for one tonne of concrete or glass - and these sorts of figures are always very approximate. Also bear in mind that embodied energy is only one area of environmental impact - all these other materials have significant impacts in a number of other areas.

7.2.2 Kiln-Drying Energy

We have no figures for the energy used in the kiln-drying of timber, although this is a widespread practice. Suffice to say, kiln-dried timber will have a higher embodied energy than air-dried or green (un-dried) timber. Oak is traditionally worked in the green state, and many Tudor building survive to attest to its longevity.

For those who wish to avoid the energy costs of kiln-dried timber, check with your supplier. You may well find kiln-dried timber to be stamped 'KD' or 'DRY'.

7.2.3 Depletion of Resources

(a) Timber Resources

Conventional wisdom has it that timber, as a 'renewable resource', can never run out. According to a standard materials reference book, "man has always had plenty of timber available for his needs....There will be changes in the availability of particular species... but there will always be sufficient timber and timber products available for constructional use."[21] The same publication also notes that, once upon a time, timber from the USA and Canada

was available in larger sections than European timber, but that this is less so nowadays, as the best and oldest timber from natural forests has already been cut.

Much timber production around the world is more akin to mining than agriculture, and like mining operations, extractive forestry is continually on the move as supplies of one species diminish until costs become prohibitive.

Country of Origin	Energy Cost of transport to UK via container ship (GJ/tonne)[20]
Papau New Guinea	2.4
Indonesia	2.2
British Colombia	1.0
Brazil	0.7
Ghana	0.6
Siberia	0.5
Finland	0.3
Sweden	0.1

Material	Embodied Energy (GJ/Tonne)
Concrete	1.0
Brick	3.1
Glass	33.1
Steel	47.5
Aluminium	97.1
Plastics	162.0

Redwoods such as African mahogany, sapele and utile, were once widely used for window frames etc, but as these become rarer and more expensive, Brazilian mahogany, meranti and lauan are taking over.[11]

In the tropics, deforestation is running at a rate that means all the commercially exploitable forests could be gone in 40 years, by about the same time as the world's oil-wells dry up,[14] and many individual species of trees are in danger of extinction now. Actual logging for the supply of timber only accounts for a small proportion of tropical forest logging (estimates vary between 5% and 30%), as most tropical timber is actually cut for firewood for domestic consumption. However, it is the timber industry that typically opens up untouched areas of tropical forest, leaving behind a network of roads that enables further clearance, often by landless farmers moving in and attempting to make a living from cleared forest.[29]

In the temperate and boreal (e.g. Siberian) regions, clear felling of old-growth forests is still the norm, and the rapidly dwindling natural forests are usually replaced with intensively managed plantations.[33] A natural capital resource with the potential to provide a permanent income is thus being replaced by an agribusiness system of doubtful sustainability.

Estimates from Friends of the Earth for the year 2010 are that the ecological capacity (sustainable supply) from the world's forests will be 2.3 billion cubic metres - yet current consumption forecasts are for 2.7 billion cubic metres.[31]

(b) Wildlife Conservation

The clear felling of old growth forests, which happens both in the tropics and in temperate and boreal regions, causes severe disruption to wildlife.[26] Reference is most often made to tropical rainforests, where possibly the greatest diversity and greatest losses have occurred. In the UK, clear felling of old growth forest is rare (old growth forest is rare itself), and the Forestry Authority has had guidelines for around 10 years now on nature conservation.[12]

(c) Soil and Water

Logging can cause disruption of ecological processes, damaging soil, causing erosion and polluting watercourses.[33] Attention is most often drawn to these effects in tropical areas, which are often seen as much the most sensitive or delicate ecosystems.

The establishment of plantation forestry can also have significant impacts on soils and water-systems. In the UK it has been reported that soils suffer erosion and drying; water courses have their flow affected, suffer sedimentation and raised levels of aluminium and other metals. Later clear felling can further exacerbate these problems.[27]

7.2.4 Global Warming

Whilst it is sometimes claimed that planting trees might be the answer to the global warming crisis, because they will 'soak up' the CO_2 from fossil fuel burning, the picture is in fact more complex.

Overall, an ecosystem such as a tropical forest, will absorb very little CO_2 - but it will have a regional cooling effect on the climate through evapotranspiration, releasing water vapour into the air and creating cloud cover.[22] Conversely, forests in far northern climates may warm the region (compared to unforested land) by absorbing heat from the sun with their dark colour.

The growing of timber itself does 'lock up' a certain amount of carbon from the atmosphere, but this must be balanced against the stress effect that logging, even selectively, has on the forest, causing further releases of carbon.[23]

In the tropics, total clearance for agriculture by burning usually follows logging, even if the logging was selective, and Greenpeace estimates that tropical deforestation (from all causes) contributes to around 18% of all global warming (30% of CO_2 emissions).[14]

In temperate regions, conversion of old-growth forest to plantation also causes a net increase in greenhouse gas levels - from release of carbon and methane in soils - as does the draining of peatlands. Plantation forestry also tends to cut trees just when they are beginning to absorb most carbon.[15]

Planting trees, where before there were none, may help absorb CO_2 and reduce global warming - and so should be part of any landscape design. But felling timber in order to build with it, however you look at it, does not.

7.2.5 Toxics

Given that modern forestry in the UK involves mainly monocrop plantations of exotic species, it is perhaps

---

UK Forestry

Total forest cover in the UK is about 10% of the land area, around 2.4 million hectares, two thirds of which is conifer forest.[7] Clear felling in state and private forests affected approximately 9000 hectares in 1990, and is expected to double in the next 20 years.[10] New planting was running at 10,900 hectares of conifer and 6,300 hectares of broadleaf trees in 1991-2. The depletion of semi-natural woodlands (wooded for more than 400 years) has now largely ended.[7]

UK grown softwood production amounted to about 800,000 cubic metres in 1990, and is expected to be double that by 2000. It consists mainly of pine and spruce, and increasingly Sitka spruce grown for construction and general purposes. About 40% of UK hardwood consumption is supplied by homegrown timber.[21]

Large areas of monocrop spruce plantations are now 40-50 years old and ready for felling - and it is hoped that this will allow a chance to pay more attention to wildlife, diversity, appearance, environment etc. in the 'second rotation'.[1]

surprising that there have been so few serious pest outbreaks. Nevertheless, it is common practice for seedling trees in the forestry nursery to be routinely treated with gamma HCH (lindane) or other insecticides.[1]

When pest infestation of forests does occur, aerial spaying with insecticides is sometimes resorted to, with, for example, organophosphorous insecticides. When this happens, very large areas usually need treating.[1] Herbicides such as 2,4-D are also widely used, for example in the suppression of heather before planting.[1]

We are not dealing with preservative treatments for timber in this chapter, but it is worth noting that more and more softwood timber is being pre-treated with toxic preservatives, possibly in an attempt to make up for inferior quality, poor seasoning and bad design.[8] (see Chapter 9 for analysis of timber preservatives)

Durability

Durability is of course a desirable quality for any building material. From an environmental point of view, if acceptable sources can be assured, then a durable hardwood is to be preferred over a softwood that requires preservative-treatment.[6]

Endangered Species

Under the 1973 Convention on International Trade in Endangered Species (CITES), to which the UK is subject, trade in the following tree species is prohibited or controlled.

Appendix I
(trade is prohibited except in special circumstances)

Alerce, (Chilean) False Larch, lahuan	(Fitztroya cupressoides, Pilgerodendron uviferum)
Brazilian Rosewood	(Dalbergia nigra)
Chilean Pine	(Aruacaria aruacaria)
Guatemalan fir	(Abies guatemalensis)
Parlatores podocarp	(Podocarpus parlatorei)

Appendix II
(trade requires special permits)

Afrormosia	(Pericopsis elata)
Ajillo	(Caryocar costaricense)
Central American/Honduran Mahogany	(Swietania humilis)
Cuban mahogany	(Swietania mahogani)
Gavilan Blanco	(Oreomunnca pterocarpa)
Quira macawood	(Platymiscium pleistachyum)

Timber species are conventionally classified into five durability ratings, from 'perishable' to 'very durable', and tables are available listing all commonly available timbers (e.g. ref. 24). Generally speaking, there are more imported, tropical timbers in the 'durable' and 'very durable' classes than there are home grown timbers. Nevertheless, species such as Cedar, Cypress, Oak and Yew are all considered 'durable', and the 'perishable' class includes European hardwoods such as beech and willow, as well as the tropical hardwood ramin.[24]

The ease with which a timber can be impregnated with preservative tends generally to be inversely related to its

durability rating.[24]

The conventional durability tests seem quite severe, involving sticking an untreated piece of timber into the ground, and waiting 5, 10, 15, 20 or 25 years for it to rot. In most building applications timber will never be exposed to such harsh conditions, and all commentators on green building stress the point that good detailing in avoiding dampness, and allowing drying out if it does occur, allows untreated, 'non-durable' timber such as fir, hemlock, pine, spruce etc. to be safely used in nearly all applications. Good seasoning, including a prior soaking in water to remove the sugars that attract fungi, is also important.[5]

7.2.6 Human Rights

According to campaigning groups, in the Americas both north and south, in the Pacific and even in Scandinavia, indigenous peoples have had their ways of life destroyed by the logging of old growth forest.[26, 33] Friends of the Earth claim that Brazilian Indians have been 'bombed, shot and poisoned' by logging companies in their hunt for mahogany.[25] Loggers are also trying to evict the Sami who have lived as nomadic reindeer herders for thousands of years in central and northern Sweden.[26] In the UK of course any evictions of indigenous peoples from forests must have occured hundreds of years ago.

There is a lack of systematic information to enable buyers to know which timbers to avoid in order to be sure of not obtaining them from sources connected with human rights abuses. But human rights issues feature strongly in the princples of responsible forestry of the FSC (see page 76).

7.2.7 Health Hazards

Bulk timber presents no hazard to workers or occupants, but wood dusts arising from working with timber may be toxic, immuno-damaging or possibly carcinogenic. Due to higher dust levels, factory workers are more at risk than those on the construction site, but dust retention measures would be a sensible precaution.[2]

Reclaimed Timber

Construction and demolition create millions of tonnes of wood waste annually. Very little of this is currently reclaimed or recycled, although commercial recycling of wood waste from other sources for pulp and particle boards is well established.

Whilst reclaimed timber is the greenest choice, there are obvious difficulties in re-using constructional timber, but there a few salvage companies in operation. Schemes built using reclaimed and recycled softwood (pitch pine) include the Birmingham Urban Wildlife Centre, the Centre for Alternative Technology's new cliff railway, and the refurbishment of Glyndebourne Opera House.[31]

Second-hand timber is often beautiful, well seasoned and saves energy costs, but it can be expensive if resawn to new sizes.[4]

The dangers of smoke from burning wood are "not significantly less than for man-made polymers".[2]

Under 1994 Occupational Exposure Limits[9] hardwood dusts have a maximum exposure limit the same as for mineral wool fibres, and are classified as respiratory sensitisers and as carcinogenic. Softwood dusts are undergoing review.

"Traditionally, [tropical hardwood] has been used as a substitute for good building detailing and design."[32]

High Impact Steel?

It has been argued that, when looking at current practice in the USA, steel might actually be a more sustainable industry than forestry. Stanley Rhodes, of Scientific Certification Systems Inc of California, a company involved in the certification of sustainable forestry operations, claimed that life cycle analysis would show steel production in US to have a lower impact than timber extraction.

It seems clear that Rhodes was attempting not so much to persuade us to build in steel, but to point up the vast difference in impacts between well- and badly-managed forestry. He describes most forestry in the US as more akin to mining than to agriculture. Clear felling affects vastly larger areas of landscape than iron-ore mining, and with current rates of extraction, old-growth forests will run out long before iron-ore deposits. He also argues that steel can be more easily recycled. But the Green Building Handbook would dispute his dismissal of the embodied-energy issue as 'insignificant', and also notes his failure to address other pollution, global warming and toxics issues. Rhodes admits properly managed sources, such as one certified by his company, are more sustainable than steel. But "as soon as wood loses its accountability, then steel can make a case for itself. The timber industry can't make it on an a priori basis any more." [18]

"Boycott Mahogany"

Brazilian mahogany, which acconts for over 10% of the UK's tropical hardwood imports, has been the centre of attention for a wide range of environmental and human rights groups, and Freinds of the Earth are now calling for a consumer boycott. They are calling for an end to the trade in mahogany until there are proper safeguards against its over-exploitation and against its illegal felling in indian reserves. Mahogany has proved impossible to grow in plantations, and only one tree per hectare is found in the wild, so huge areas of forest are affected by its extraction. According to Friends of the Earth there are no longer any big or reliable stands of mahogany, and not enough young trees to ensure regrowth.

The logging of mahogany on indian reserves is of particular concern. Despite being illegal under Brazilian law, it is known to continue. Although there has been an agreement between the timber exporting and importing industries since 1992 (the TTF-AIMEX accord) not to deal with illegally felled mahogany, this is limited to only one state in Brazil (Para). And FoE claim to have documentary evidence that AIMEX members were still logging illegally in 1993-4.[36]

Mahogany Fences?

A protest group called CRISP-O (Citizen's Recovery of Indigenous People's Stolen Property Organisation) has been encouraging 'ethical shoplifting' of mahogany. They claim that much mahogany imported to the UK from Brazil is illegally felled. Supporters have been removing mahogany from stores and timber yards, and handing it in to the authorities as stolen property.[19]

"My most popular building designs, with few exceptions, have been wood framed. The rougher and more massive the timbers the more appealing the designs."
Malcolm Wells [17]

Woods of the World

Woods of the World is an interactive computer database that provides a wide range of data for over 900 species of tropical and temperate woods. Information includes physical, mechanical and woodworking characteristics, as well as an environmental rating (from 'not-threatened' to 'extinct' on a ten point scale) and lots of colour pictures. The main function of the program is to enable users to choose alternative lesser-known species with the same properties as overused or threatened ones. (See Dare to be Different above.)

The database includes details of timbers from well-managed sources as certified by FSC accredited bodies, (see page 76) and also lists of suppliers etc. It comes from the United States, where there are somewhat more certified-source timbers available, and though we haven't seen it in action yet, we expect it is somewhat biased towards that part of the world.

You will need a PC or Mac, with a CD-ROM and/or lots of hard-disk space, and $264. Contact: Tree Talk Inc., PO Box 426, Burlington, VT 05402 USA. (tel: 001-802-863-6789 fax: 001-802-863-4344)

Less is More with the new Eurocode

The European standards body CEN has introduced DDEN1995-1-1 Eurocode 5, Design of Timber Structures. In force from December 1994, for a trial 5 year period, it runs in parallel with the current BS 5268, which governs the structural use of timber. Designers may use whichever they prefer - but according to Friends of the Earth, the Eurocode allows a saving of 10-20% in timber used by, for example, allowing wider spacing of timbers in stud walls.

7.3 Best Buys

(a) Buy Certified Timber if Available

Timber certified to FSC standards will clearly be a Green Building Digest Best Buy. But reliable independent certification schemes such as these are only just in their infancy, and they are unlikely to be sufficiently well developed in the near future to fully meet demand for sustainable timber. So if you can't get certified timber, the next best is to ask a 1995 Group company if they have products meeting your environmental requirements - they should have information to back up their claims.

Apart from that, the old advice still holds true when buying timber - ask where the timber has come from and what assurances the seller can give about its environmental credentials. Ask for certified timber, and if the supplier doesn't know what it means, tell them about it or suggest they get in touch with a certifier (see listings), the FSC or the 1995 Group.

(b) Beware of bogus claims

Beware of claims about sustainable sources that are not certified. The claim that for every tree felled so many are planted, for example, could involve felling virgin forests and creating plantations.

(c) Buy Locally - especially if suitable certified timber is not available

Buying locally-sourced timber not only saves on transport energy costs, but also coincides quite well with our assessment of general forestry practice - home-grown timber production, especially from some of the small suppliers listed here, is generally reasonably well-managed even if not certified. As with all green purchasing, the more local the better, but often 'home-grown' is the best that can be managed. Buying more locally is not always easy, and may involve contacting woodland owners or sawmills rather than conventional building timber suppliers. The FICGB's Woodmark is also widely used as an indication of British grown timber.

With imported timbers, Scandinavian timber has least far to come, and so is the next best for UK buyers. It has also been suggested that replanting is perhaps taken more seriously in Scandinavia, whilst logging of old-growth forest is more prevalent in America,[4] but this may be unprovable.

Much tropical timber is not only associated with the worst forestry practices, but has furthest to come.

(d) Reduce Consumption

As with any other material, specifiers might also make a point of reducing consumption. Reduced consumption of timber should obviously not lead to increased demand for other materials, unless the impact of the new material is known to be less than that of timber. Techniques for reducing consumption include using reclaimed timber, using smaller sections, ensuring that design and installation are conducive to durability and designing for minimal wastage. (For example, specifying finished planed sizes to be just smaller than standard sawn sizes is less wasteful than specifying planed timber at conventional sawn sizes, which will have to be planed down from the next larger standard size.)

(e) Dare to be Different?

Specifying unusual timber, either unusual species or unusual sizes, signals a demand for diversity to forest management. With homegrown timbers, specifying large sections encourages well-managed, thinned plantations, whilst conversely, specifying smaller than usual sections adds value to thinnings which might otherwise go to waste or firewood.[4]

It is also argued that by using the full range of available species from tropical forests, deforestation there will be slowed. Present logging practice focuses on just a few well-known species, and by using more timber from each unit area logged, the total area of forest cleared should be reduced.[13] (See Woods of the World below.)

Some doubts have been raised about this approach though. It is possible that cutting more species would lead to a more complete clearance, which harms natural forest regeneration. There is also concern about exploiting species about which little is known in terms of conservation status, distribution and ecological requirements.[29]

(f) Choose Your Supplier

There are a large number of smaller environmentally conscious suppliers with a wide range of timbers to choose from. There are also members of the WWF's 1995 Group, who are at least making a commitment to move towards well-managed supplies (see page 79), and who should stock at least some certified timber as well as know about their other sources.

"Many developing countries question our right to advise them on the use of their forest resources, when forests in developed countries are not treated with any greater respect. Now is the time for both the developed and the developing world to realise they have a mutual responsibility for the world's forests." [26]

7.4 Specialist Suppliers

Capel Iago Sawmills, Llanbydder, Dyfed SA40 9RD (tel: 01570 480464)

Contact: John Stephens

Product: Local Hardwoods & Reclaimed Timber

Locally grown hardwoods and softwoods. Specialist in the supply of reclaimed timbers especially pitch pine and oak which is marketed in beam sections or as flooring. Some reclaimed tropical hardwood such as jarrah is handled but we are assured that new timber of this origin is very much avoided. Capel Iago Sawmills prefer to source their new timbers from storm damage or from sustainably managed local woodlands in Dyfed. A mobile sawmill is also operated which allows customers' timbers to be converted at point of use.

Other Products:Flooring

Carpenter Oak & Woodland Ltd, Hall Farm, Thickwood Lane, Colerne, Chippenham, Wilts, SN14 8BE (tel: 01225 743089, fax: 01225 744100)

Contact: Charles Brentnall

Product: Oak & Chestnut Timber

Suppliers of oak (green or seasoned) in beam sections or as other traditional building craft consumables, such as; hand cleft carpenters pegs (for timber framing) and plaster lath in oak or sweet chestnut. They also make oak roofing shingles cleft from english and oak tile pegs.

The company has a replanting policy whereby the plant three trees for every one they use.

Other Products:Green Oak Buildings / Oak Shingles / Traditional Roof Structures

Chadzy's Salvage, Bryneithyn, Aberarad, Newcastle Emlyn, Dyfed (tel: 01239 710799)

Contact: Chadzy

Product: Salvaged Timbers

Suppliers of reclaimed softwoods including pitch pine and floorboards. Also suppliers of old 'unused' stock timbers of many various sizes.

Other Products:Reclaimed building materials

Crendon Timber Engineering Ltd, Drake's Drive, Long Crendon, Ayelsbury, Buckinghamshire, HP18 9BA (tel: 01844 201020 fax: 01844 201625)

Product: PARALLAM

Reconstituted timber trusses, joists, beams and purlins. Made from second growth Southern pine from British Columbia in Canada. This product provides an alternative to glulam beams but is not as adaptable for large spans. Parallam is only available in lengths up to six metres.

The adhesive content is higher than that contained in glulam beams because of the much smaller timber pieces incorporated.

Dartington Hall Trust, Elmhirst Centre, Dartington, Totnes, Devon TQ6 EL (tel: 01803 866688)

Contact: Charles taylor

Product: Douglas Fir & Larch

Dartington Hall trust is the first timber grower in the UK to receive the Soil Association WOODMARK accredation. for its sustainable production of douglas fir and larch timber.

Ecological Trading Company (1989) Ltd, 659 Newark Road, Lincoln LN6 8SA (tel: 01522 501 850 fax: 01522 501 841)

Contact: John Ward

Product: Timber Importers

Importers of sustainably produced tropical hardwoods. The ETC trades directly with the producers of the timber rather than with agents. They can therefore monitor the extraction themselves. Also as trade is direct the producers receive higher prices for their lumber on condition that they continue to operate in an environmentally acceptable manner. One of the central goals of the ETC is to assist communities in the development of sustainable forestry projects aimed at a general improvement in living standards and care for the environment.

ETC are presently investigating the possibilities of establishing a window/ door manufacturing service using their sustainably produced tropical timbers. If this goes ahead it will provide a unique service within a market which is, at present, one of the most difficult to obtain 'environmentally' acceptable goods.

Glue Laminated Timber Association, Chiltern House, Stocking Lane, Hughendon Valley, High Wycombe, Buckinghamshire HP14 4ND (tel: 01494 565180 fax: 01494 565487)

Glue laminated timber association representing over eleven producers of these products. Contact them for your nearest producer.

Glulam producers use only redwood or European whitewood (picea abies) from Scandinavia. The adhesive used is phenol-resorcinol-formaldehyde (PRF)

Home Grown Hardwoods, Goginan, Aberystwyth, Dyfed SY23 3NP (tel: 01970 880294)

Contact: Alun Grifiths

Product: Welsh Hardwoods

Merchant in the West Wales area for homegrown hardwoods. Machining service available. Specialise in the provision of wide board flooring. As the name suggests 'Home Grown Hardwoods' specifically avoid the use and sale of tropical hardwoods of any species.

Other Products:Joinery and Flooring

J.R.Nelson, Dupree Partnership Ltd, The Sawmill, Wills Farm, Newchurch, Romney Marsh, Kent, TN29 0DT (tel: 01233 733361)

Contact: Jeremy Nelson

Product: Reclaimed Pitch Pine

Specialist suppliers of reclaimed pitch pine, including milling & fabrication service.

Three grades of reclaimed pitch pine are available; 'A' 'AB' & 'B'. As they specialise in pitch pine this supplier has extensive stocks of this timber.

Other Products:Reclaimed Pitch Pine Windows & Doors, Stairs and spindles

Marlwood Ltd, Court Wood Farm, Forge Lane, East Farleigh, Maidstone, Kent. (tel: 01622 728718 fax: 01622 728720)

Contact: Keith Elliment

Product: Woodmiser Bandsaw Agents

Marlwood are distributors of various mobile bandsaw types and are able to put you in touch with a sawyer in your area. Mobile bandsawing is a very positive way of ensuring that your timber requirements come from a local source. By using mobile sawmills you are ensuring that a value is placed on local woodlands that otherwise might be neglected. Most sawyers have access to kilns, if kiln dried local timber is required.

Milland Fine Timber Ltd, The Sawmill, Iping Road, Milland, Nr Liphook, Hampshire GU30 7NA

Contact: Charles Townsend

Tel: 01428 741505

Fax: 01428 741679

Product: Distributors of sustainably produced tropical hardwoods for the Ecological Timber Company. Milland also market locally produced and European hardwoods and hardwoods from the Minominee tribe sustainable forest reserve of over 220,00 acres in Wisconsin, USA. The prime timber from this supply is; maple, red oak, yellow birch and white pine. This is the first large scale North American forestry operation to be certified by an independant body- Scientific Certification Systems (SCS).

Prencraft, Garnbwll, Mynyddcerrig, Pontyberem, Llanelli, P Carmarthenshire, SA15 5BN (tel: 01269 870031)

Contact: Arwyn Morgan

Product: Locally Produced Hardwoods

Locally grown hardwoods and softwoods only.

Ridgeway Timber, Cwmbrandy Gardens, Manorowen, Fishguard, Pembrookshire, SA65 9PT (tel: 01348 873179 Mobile: 01378 147300)

Contact: Stephen Cull

Product: Woodmiser Portable Sawmill

A cost effective mobile sawmilling service throughout Dyfed and West Wales using a woodmiser portable mill. Able to convert on-farm timber for a multitude of uses including building and fencing. Ideal for converting green oak beams for cottage and barn conversions. Mobile sawmilling is an environmentally sensitive method of providing local timber for local uses. It also has the advantages of encouraging commercial and sustainable use of small 'neglected' woodland and providing a source of local employment.

Ronson Reclamation, Norton Farm, Wainlodes Lane, Norton, Gloucestershire GL2 9LN (tel: 01452 731236)

Contact: Ron

Product: Reclaimed and Salvaged Timber

Suppliers of reclaimed and salvaged timber including oak and elm.

Thompsons Sawmill, Slugwash Lane, Wivelsfield Green, East Sussex, RH17 7SS (tel: 01444 454554)

Contact: Richard Thomlinson

Buys either windblown or sustainably managed standing timber from small woodland owners in the region. Offers for sale all native hardwoods and home grown softwoods. Freshly sawn or air/kiln dried.

Treerights, Westerton cottage, Killearn, Sterlingshire GB3 9RT

(tel: 01360 550873)

Contact: Nick Pye

Product: Local Hardwoods & Softwoods

Locally harvested timbers using a mobile sawmilling machine.

Other products:Green oak frame buildings.

Treework Services Ltd, Cheston CombeChurchtown, Blackwell, Near Bristol BS19 3JQ (tel: 01275 464466 fax: 01275 463078)

Contact: John Emery

Product: European & Homegrown Hardwoods

Suppliers of sustainable timbers of UK or European origin.

Treework Services Ltd offer a comprehensive range of services to timber purchasers and also woodland owners.

The company has an ecological policy which restricts it to the marketing of only sustainably produced timbers. The Company therefore supplyies homegrown, European and North American hardwoods and specialist softwoods.

The company has pledged a percentage of its profits to replanting schemes at home and abroad. It was instrumental in establishing Tree Aid - now a major charity within the forestry and timber industry to assist, through tree planting, the stricken Sahel region of Africa

Whipple Tree Hardwoods, Milestone farm, Barley road, Flint Cross, Near Heydon, Royston, Herts SG8 7QD (tel: 01763 208966)

Contact: Hugh Smart

Product: Homegrown Timbers

Specialists in English oak for the conservation and restoration of older buildings. Oak, elm and sweet chestnut timbers available. Full range of English oak air and kiln dried for high class joinery and furniture.

Other Products:Hardwood flooring.

Witney Sawmills, The Old Vicarage, Clifford, Hereford, Hereford & Worcester, HR3 5EY (tel: 01497 831656 fax: 01497 831404)

Contact: Willy Bullow

Product: Homegrown timbers

Witney sawmills have an environmental policy which restricts marketing to only locally grown hardwoods. Mainly oak and selected softwoods (douglas fir, larch and cedar).

The company also has a policy of woodland creation and management whereby they plant more trees than they use.

The Woodshed, Battle Bridge Centre, 2-6 Battlebridge Road London NW1 2TL (tel: 0171 278 7172)

Contact: Crispin Mayfield

Reclaimed and antique pine in various widths and lengths. Sold in two grades 'A' & 'B'. Perfect for the DIY'er as lengths under 2 metres are sold at half price.

Other Products:Paint Supplies

N.B. Claims for a product's sustainability in this listing are the suppliers' own, as supplied to the GreenPro database.

7.5 Certification Organisations

7.5.1 The Forest Stewardship Council

The FSC is an international body set up by environmentalists, foresters, timber traders, indigenous peoples' organisations, community forestry groups and certification organisations from 25 countries. Its main function is to evaluate, accredit and monitor certification organisations such as the Soil Association, in order to guarantee the authenticity of their claims. Its principles of forest management are designed to ensure good forest stewardship in environmental, social and economic terms, and it has the support of environmental groups such as WWF and FoE.

FSC Headquarters: Dr Timothy Synott, Executive Director, Forest Stewardship Council, Avenida Hidalgo 502, Oaxaca 68000, Oaxaca, Mexico tel: 0052 951 46905, fax 0052 951 62110.

UK FSC Coordinator (Hannah Scrase), Oleuffynon, Old Hall, Llanidloes, Powys SY18 6PJ tel/fax: 01686 412176.

Listing on previous page supplied by the Green Building Press, extracted from 'GreenPro' the building products and services for greener specification database. At present GreenPro lists over 600 environmental choice building products and services available throughout the UK and is growing in size daily. The database is produced in colaboration with the Association for Environment-Conscious Building (AECB).

For more information on access to this database contact Keith Hall on 01559 370908
or email buildgreen@aol.com

(a) FSC Forest Stewardship Principles

1.   Compliance with Laws and FSC Principles
Forest management operations shall respect all applicable laws of the country in which they occur, and international treaties and agreements to which the country is a signatory, and comply with all FSC Principles and Criteria.

2.   Tenure and Use Rights and Responsibilities
Long-term tenure and use rights to the land and forest resources shall be clearly defined, documented, and legally established.

3.   Indigenous' Peoples Rights
The legal and customary rights of indigenous peoples to own, use and manage their lands, territories and resources shall be recognised and respected.

4.   Community Relations and Workers' Rights
Forest management operations shall maintain or enhance the long-term social and economic well-being of forest workers and local communities.

5.   Benefits from the Forest
Forest management operations shall encourage the efficient use of the forests' multiple products and services to ensure economic viability and a wide range of environmental and social benefits.

6.   Environmental Impact
Forest management shall conserve biological diversity and its associated values, water resources, soils, and unique and fragile ecosystems and landscapes, and by so doing, maintain the ecological functions and integrity of the forest.

7.   Management Plan
A Management Plan - appropriate to the scale and intensity of the operations - shall be written, implemented and kept up-to-date. The long term objectives of management, and the means of achieving them, shall be clearly stated.

8.   Monitoring and Assessment
Monitoring shall be conducted - appropriate to the scale and intensity of forest management - to assess the condition of the forest, yields of forest products, chain of custody, management activities and their social and environmental impacts.

9.   Maintenance of Natural Forests
Primary forests, well-developed secondary forests and sites of major environmental, social or cultural significance shall be conserved. Such areas shall not be replaced by tree plantations or other land uses.

10.   Plantations *(draft principle not yet ratified)*
Plantations shall complement, not replace, natural forests. Plantations should reduce pressure on natural forests.

7.5.2 Other Certification Organisations

Scientific Certification Systems Ltd (SCS), 1611 Telegraph Avenue, Suite 1111, Oakland, California 94612

Contact: Debbie Hammel

(tel: 001 510 832 1415 fax: 001 510 832 0359)

SCS offer an international sustainable forestry (management and harvesting) certification system which is currently being assessed for FSC accreditation. News due in July 1995.

Other Products:

SGS Forestry, Oxford Centre for Innovation, Mill Street, Oxford, OX2 0GX (tel: 01865 202345 fax: 01865 790441)

Contact: Frank Miller

SGS Forestry offer an international sustainable forestry (management and harvesting) certification system which is currently being assessed for FSC accreditation. News due in July 1995.

Smart Wood (Rainforest Alliance)

Contact: Richard Donovan

(tel: 001 802 899 1383 fax: 001 802 899 2018)

SMART WOOD offer an international sustainable forestry (management and harvesting) certification system which is currently being assessed for FSC accreditation. News due in July 1995.

The Soil Association, 86 Colston Street, Bristol BS1 5BB (tel: 0117 9290661 fax: 0117 9252504)

Contact: Dorothy Jackson, Ian Rowland or Rod Nelson

The Soil Association's WOODMARK scheme aims to promote good forest management worldwide through reliable certification and labelling of wood from well-managed sources.

The scheme began in 1992 after the UK Timber Certification Working Group (comprising environmental groups, foresters and timber traders) encouraged the Soil Association to apply its philosophy of sustainable resource use and its certification expertise to forests and timber.

WOODMARK will operate under the auspices of the Forest Stewardship Council, the international organisation which will monitor the activities of accredited users.

The Soil Association is a registered charity founded in 1946, which exists to research, develop and promote sustainable relationships between soil, plants, animals, people and the biosphere, in order to produce healthy food and other products while protecting and enhancing the environment.

Other Products: Organic food certification scheme

7.6.3 Certified Operations

As yet, there are only 17 certified timber sources, and the suppliers from these sources are mostly in the USA. There are, however, a lot more 'in the pipeline'. The following is a list of all sources certified by organisations (expected to be) accredited by the FSC, from a survey by WWF.

Name of Operation	Certifying Organisation	Products	Area of Woodland	Availability in UK
AMACOL Ltda, Portel, Para, Brazil	Rainforest Alliance	plywood & veneers	59,000 ha	
Bainings Community-Based Ecoforestry Project, Rabaul, Papua New Guinea	SGS Silviconsult Ltd	Mixed 'Redwood' mouldings (taun)	12,500 ha	B & Q plc
Chindwell Doors, Johor, Malaysia	SGS Silviconsult Ltd	Rubberwood Doors	3,284 ha	B & Q plc
Collins Pine, Chester, California	Scientific Certification Systems	Timber (hem fir, pine etc.)	38,300 ha	
Collins Pine, Kane Hardwood, Pennsylvania	Scientific Certification Systems	Veneer, Flooring (cherry, oak, maple etc.)	48,300	
Dartington Home Woods, Devon	Soil Association	Sawlogs & Bars (fir, larch, thuja etc.)	92 ha	Silvanus, 15 Link House, Tiverton, Devon EX16 5LG tel: 01884-257-344
Demerara Timbers Ltd, Guyana	SGS Silviconsult Ltd (Certified to DTL's own criteria, not the FSC's)	Greenheart, Purpleheart, Demerara mahogany etc.	500,000 ha	
Keweenaw Land Association Ltd, USA	Rainforest Alliance	Timber & Veneer (maples, basswood, birch etc.)	50,000 ha	
Masurina, Papua New Guinea	Rainforest Alliance	Timber (Dillenia, Kwila, Mersawa etc.)	5,000 ha	Ecological Trading Company Ltd.
Menominee Tribal Enterprises, USA	Scientific Certification Systems	Timber (Pine, Hemlock, Maples, Birch, Oak etc.)	97,500 ha	Milland Fine Timber Ltd.
Pengelli Forest, Wales	Soil Association	Chairs (Oak, Ash) (Timber for own use only)	65 ha	Dyfed Wildlife Trust, 7 Market St, Haverford West SA61 1NP tel: 01437-765-462
Perum Perhutani - State Forestry Corp, Indonesia	Rainforest Alliance	Teak, Mahogany, Rosewood, Pine etc.	2,063,100 ha	
Pingree Family, Seven Islands Land Man. Co., USA	Scientific Certification Systems	Spruce, Fir, Birch, oak etc.	406,250 ha	
Plan Piloto, Mexico	Rainforest Alliance	amapola, k'atalox, bari, machich etc.	95,000 ha	
Plan Piloto Forestal, Honduras	Scientific Certification Systems	Honduran Mahogany, Cedar etc.	33,000 ha	
Portico SA	Scientific Certification Systems	Doors (Carapa, Gavilan)	3,900 ha	
Tropical American Tree Farms, USA	Rainforest Alliance	not yet available	595 ha	

7.6 Sustainable Initiatives

7.6.1 Local Authorities

The Local Authorities Project of the Soil Association's Responsible Forestry Programme is an initiative to deliver an information service on the formulation and implementation of 'good wood' policy to Authorities throughout Britain.

The Local Authorities Project is actively seeking and collating information from Authorities on the policies which they now have, and the degree of implementation which they are able to achieve.

It is clear that there are a large number of authorities who have adopted policies which eschew the use of tropical hardwood except under very particular circumstances. We feel that whilst these policies represent a display of good intent towards the environment, that there are flaws which should be addressed. They do, for instance, discriminate against tropical producers who are investing in responsible forestry practice. They also fail to discriminate against temperate and boreal producers who are indulging in dubious forestry practice.

The extent to which Authorities implement the 'good wood' policies which they have adopted seems to vary enormously. Some Councils appear to have an extraordinarily casual attitude towards their own policy implementation, and others are vigorous in their commitments, with all shades in between.

The Local Authorities Project will be of assistance to all local authorities who wish to evolve their policies towards a more rational and better informed environmental standpoint on timber purchasing. It will advise those who are charged with the implementation and monitoring of these policies in a positive and informed way, and it will continue to gather information on a nationwide basis.

For more information contact: Rod Nelson, Research and Information Officer, Responsible Forestry Programme, The Soil Association, 86 Colston Street, Bristol BS1 5BB tel: 0117 929 0661 fax: 0117 925 2504.

7.6.2 Wood Lots

The Forestry Authority is publishing 'Wood Lots', an exchange and mart offering free advertisements for anyone wishing to sell domestic timber, from a few planks, to whole woods. Initially it is being run as a pilot in the south east (The Weald) in association with East Sussex County Council, and in the north west (Cumbria and Lancs). Contact Vince Thurkettle, Market Development Officer, The Forestry Authority, Great Eastern House, Tenison Rd., Cambridge CB1 2DU tel: 01223-314546 for details.

7.6.3 Coed Cymru

The appearance of a large number of environmentally conscious suppliers of Welsh timber is largely thanks to the work of Coed Cymru. This organisation is fostering hardwood production and marketing in Wales - including hardwood for construction purposes. It encourages woodland management practices such as maintaining continuous cover, selective felling, natural regeneration and playing with light to promote good growth. It claims that this is not only environmentally popular, but primarily done to produce good wood at a low management cost.[34] Contact: Coed Cymru, 23 Frolic Street, Newtown, Powys SY16 1AP tel: 01686-628514.

7.6.4 The WWF 1995 Group

The WWF 1995 Group is a partnership between the World Wide Fund for Nature and over 40 UK companies which are 'determined to play their part in improving the quality of forest management worldwide'. It was set up in 1991 with a call for the international trade in wood and wood products to be based on well-managed forests by the end of 1995. Members are committed to the FSC as the only currently credible labelling scheme.

*They have to meet targets for a) knowing **where** all their timber sources come from, b) then using their best judgement and independent advice to only buy from well-managed sources, c) having at least some independently certified timber in their supplies by 1995, and d) working towards all timber being independently certified by 1999. These targets should mean members are using or supplying an increasing proportion of FSC certified sources and a decreasing proportion of self declared well-managed sources. The following is a partial list of members - those that might be of interest to our readers. (Others include some major retailers and paper users.)*

Members as at June 1995	
Acrimo Ltd	M & N Norman
B & Q plc	Magnet Ltd
Bernstein Group plc	MFI
Chindwell Co Ltd	Milland Fine Timber
Core Products Ltd	Moores of Stalham
Crosby Sarek Ltd	(UK) Ltd
Do It All ltd	Premium Timber
Ecological Trading	Products Ltd
Company	Rectella International Ltd
F R Shadbolt & Sons Ltd	Richard Burbridge Ltd
F W Mason & Sons Ltd	Richard Graefe Ltd
Great Mills (Retail) Ltd	Spur Shelving
Harrison Drape	Swish Products Ltd
Helix Lighting	Texas Homecare Ltd
Homebase DIY	Wickes Building
Laing Homes Ltd	Supplies Ltd
Larch-Lap Ltd	Woodbridge Timber Ltd
	Woodlam Products

Composite Boards 8

8.1 Scope of this Chapter

This chapter looks at the environmental impacts of the major composite board products available on the market, including Plywood, Chipboard, Fibreboards, Oriented Strand Boards and Cement-Bonded Boards. The alternatives section looks at straw particle boards, stramit

and Tectan - a board manufactured from recycled tetrapak packaging.

The report does not cover wood preservatives, fire retardants or insecticides used in composite boards. Preservatives are covered in Chapter 9.

8.2 Introduction

8.2.1 Types of Board

(a) Plywood & Blockboard

Plywood is manufactured from thin wood veneers, glued together into boards, generally using formaldehyde or occasionally, isocyanate resins. The veneers are produced by soaking logs to soften the fibres, then peeling the veneer off using a rotary cutter or by slicing.[7] The plywood 'core' may consist of particleboard, hardboard or other materials, rather than a veneer.[40]

Blockboard and laminboards are composite boards with core made up of strips of wood about 25mm wide (blockboard) or veneer (laminboard), laid separately or glued or otherwise joined together to form a slab, to each face of which is glued one or more veneers.[1]

(b) Particleboards

Particleboards are made from wood flakes, chips or fibres,[1,9] using virgin logs or 'waste' wood such as offcuts from timber mills, forest thinnings and undersized stock[32] as the raw material. Recycled wood currently only accounts for a very small proportion of particleboard manufacture, but there is increasing use of broken pallets and discarded timber recovered from landfill sites.[7, 33] The major source of wood for chipboard, fibreboard and Oriented Strand Board is plantation grown softwood.[7, 33]

Chipboard is a particleboard made from small particles of wood and a synthetic resin binder. Chipboard typically contains 7-10% resin,[1, 8] usually urea-formaldehyde, or melamine urea-formaldehyde for improved moisture resistance boards.[7]

Currently, 60% of chipboard used in the UK is imported.[1] Chipboards are graded C1 - C5, C5 being the most durable in damp conditions up to 85% humidity.[7] No grade of chipboard is suitable for permanently wet conditions.[7]

Oriented Strand Board is made from wood strands, oriented to simulate some of the characteristics of three-ply plywood.[1]

(c) Fibreboards

Fibreboards can be manufactured from a number of materials, including wood pulp, flax and sugar cane.[22] although wood is by far the most common raw material. Boards are constructed by mechanically breaking down solid wood into fibres which are felted and then reconstituted under heat and pressure. Fibreboards do not usually include a resin binder, the primary bond usually derived from the inherent adhesive properties of the felted fibres.[1] The exception to this is medium density fibreboard (MDF), which includes a bonding agent. It may be wise to check with the manufacturer if a particular product contains resins or binders.

MDF is manufactured by a similar process to other fibreboards except that the primary bond within the board is achieved using a bonding agent,[1,9] usually urea formaldehyde.[7]

(d) Cement Bound Boards

Wood-cement particleboard and wood-wool cement slabs comprise 25% wood chips or strands by weight, the remainder made up by a binder of ordinary Portland cement.[9] Small amounts of chemicals may also be used to accelerate cement setting.[7]

(e) Melamine Finishes

Chipboard and plywoods are available with melamine surface finishes.[38] These consist of plain or decorative surface paper impregnated with melamine formaldhyde resin and consolidated under heat and pressure, with piles of core paper permeated with phenol formaldehyde resin.[38] For melamine laminated products, we recommend the impacts of the laminate be added to that of the board. However, the additional impacts of melamine laminates could be considered to be offset somewhat by the increased durability which they impart, and the displacement of a need for paint or varnish.

8.2.2 Environmental Benefits of Composite Boards

Board products have some environmental advantages over sawn lumber, namely that the use of wood is more economic than sawn lumber, and softwoods, recycled material and other plant based products can be used in their manufacture.[7] They are also usually cheaper and can have structural advantages.

8.2.3 Environmental Concerns

(a) Energy Use

Composite boards use far more energy to manufacture than sawn wood, in the production of resins (which are usually petrochemical based) and cement, wood preparation and in the heat and pressure processes used to form the board.

(b) Resource Use

A second concern is the use of tropical hardwoods and consequent rainforest destruction, particularly in plywoods imported from Asia and South-East Asia. High durability plywoods and decorative veneers are the products most likely to be made from tropical hardwoods[(8)] and the plywood industry plays a significant role in rainforest destruction. For example, uncontrolled exploitation of Ceibas pentandra for plywood at Iquitos, in the Peruvian Amazon during the 1970s led to the total depletion of the resource in that area.[11] The shortage of hardwood supply is reflected by rising cost, with 1995 prices 25% higher than 1994.[10]

In 1994, the UK imported 1 million cubic metres of plywood,[10] nearly three quarters of which originated from South East Asian countries[53] - mainly Indonesia and Malaysia, the worlds largest exporters of plywood.[51, 52]

An advantage to third world countries exporting timber based boards is that the product is manufactured in the country of origin, thus generating more profit for the country than would be gained from exporting raw wood. However, this is often at the expense of destructive forestry practices, and large amounts of energy required to transport the product to the UK.

(c) Resin Chemistry

Off-gassing of resin chemicals such as formaldehyde is also of concern with regard to the health of workers and building users -although the industry is taking steps to minimise this risk, and zero-formaldehyde boards are available.[7] (See alternatives section, p.94).

The two major adhesive groups used in the manufacture of timber based boards are iso-cyanate and formaldehyde based synthetic resins. The formulation used depends on the required durability of the board as the resins not only bind the wood members but also impart properties that extend the life and range of the board.[32]

Urea formaldehyde (UF) is the most frequently used, but suitable only for dry conditions. Phenol- and Melamine-formaldeyde (PF & MF) resins have improved moisture resistance and are used in 'weather and boil resistant' (WBR) boards, either singly or in combination with urea formaldehyde. Resorcinol formaldehyde (RF) is used in the manufacture of weather and boil proof (WBP) boards.[9]

Resin binders in composite boards have been found to yield measurable amounts of formaldehyde, particularly when the board has not been treated with an impermeable surface. Measurements in buildings are far lower than those found in industry, but there is concern over the long hours of exposure in the domestic environment.[20] For use in poorly ventilated areas or in a bedroom, it may for health reasons be worth finding an alternative board which doesn't contain formaldehyde.[22]

Formaldehyde emissions have been linked to sick building syndrome (discussed in chapter 15). Possible health effects include respiratory problems, locomotive disorders,[17] dermatitis, rashes and other skin diseases,[16, 20] Formaldehyde is classed as an animal carcinogen and probable human carcinogen.[20]

Phenol formaldehyde particularly has been linked to dermatitis, rashes and other skin diseases[16, 20] as well as respiratory complaints associated with exposure to component vapours.[20] Occupational exposure to synthetic glues based on Carbamine- and Phenol-formaldehyde resins have been linked to catarrhal respiratory disease and locomotive disorders.[17] The British timber industry has taken steps to minimise off-gassing of formaldehyde and standards are being set which appear to minimise any health risk.[32] BS5669 sets the limit for formaldehyde content for particleboards at 25mg per100g of board; (German standards set a maximum of 10mg per 100g) and Medium-density fibreboard is covered by BS1142. Board which satisfies any of these standards "would not normally cause any irritation".[35]

Isocyanate resins are increasingly used as non-offgassing replacements for formaldehyde based resins.[32] These are also more efficient, enabling a higher bond strength with a smaller resin content,[9] although they can cause skin irritation[7, 32] and have been linked to Reactive Airways Dysfunction Syndrome (RADS) in workers exposed to high doses during its manufacture.[25,32,34] The symptoms, similar to asthma, can be brought on by a single exposure[26] and once sensitised, exposure to even extremely low doses can lead to a severely disabling reaction.[20] The use of isocyanate resins is limited mainly by its relatively high cost,[35] and the requirement for much more careful handling during resin- and board-manufacture.[7] When burned, diisocyanate resins give off toxic hydrogen cyanide fumes.

Boards using 'natural' glues are also available. Glues manufactured from soya, blood albumen, casein and animal products have a lower toxicity than their synthetic counterparts and are not derived from petrochemicals, but still require large amounts of energy in their manufacture.[42] Natural glues are only suitable for internal use and so their application will be more limited.[1]

Table 1:

Popular Uses of the major board types

	Plywood			Chipboard			Fibreboards			
	WBP Bonded Ply	Non-WBP Bonded Ply	Lamin- & Blockboard	Type I	Type II	Type III & II/III	Hardboard	Mediumboard	Softboard	MDF
Sheathing	x					x	x	x	(x)	
Flat Roofing	x					x			(x)	
Roof Sarking	x					x			(x)	
Cladding	x							x		
Floor Underlay							x	x	x	
Floor Surface (Dry)	x				x					x
Floor surface (moisture hazard)	x					x				
Linings: Interior partitions & wall panels, ceilings & roofs	x	x		x			x	x	x	
Structural Components										
Composite Beams	x						x			
Truss Gussets	x									
Stressed skin floor & roof panels	x				x	x	x			
Joinery etc.										
Fascias & Soffits	x							x		
Staircase construction	x		x		x	x				x
Window joinery										x
Mouldings & architraves				x	x					x
Furniture		x	x	x			x			x
Doors	x	x		x			x		x	x
Temporary works										
Concrete formwork	x					x	x			
Signs & hoardings	x					x	x		x	
Shopfitting & display work.	x	x	x	x			x	x		x

(x) = Bitumen Impregnated Softboard

8.3 Best Buys

Due to the numerous applications and different board specifications, table 1 has been included to assist in finding the 'greenest' board for a particular application.

(a) Occupant Health

In terms of occupant health, the use of non-resin boards such as soft- medium- or hardboard, or cement bonded board are the 'healthiest' boards. In applications where these are not suitable, ply- or particleboards bonded with diisocyanate resins are considered to be non-offgassing, although their environmental impact is high in terms of resin manufacture. Of the formaldehyde resins, phenol formaldehyde is the most stable and has a much smaller offgassing problem than urea-formaldehyde.

(b) Resource Use

In terms of resource use, best buys are particle or fibreboards as these can be manufactured using low grade or 'waste' wood, whereas plywood requires the use of whole logs.

(c) Alternatives to Hardwood Ply

The use of hardwood plywood can be difficult to avoid for high specification products, although mediumboard has been recommended as a tried and tested alternative to plywood for structural sheathing in timber houses, due to its high racking strength. It is not suitable as an alternative material where high moisture resistance, bending or tension strength are required.[54]

Working With Wood Products

It is recommended that dust retention measures and respiratory protection are employed on both the construction site and in the manufacturing plant, when working with wood based board products.[20]
Hardwood dusts have a maximum exposure limit the same as for mineral wool fibres and are classified as carcinogenic and respiratory sensitisers. Softwood dust exposure limits are currently under review.[48] Cutting or sanding chipboard (and, presumably, other formaldehyde resin bonded board products) will also temporarily increase the rate at which formaldehyde is released.[21]

	£	Production									Use					ALERT!
	Unit Price Multiplier	Energy Use	Resource Depletion (bio)	Resource Depletion (non-bio)	Global Warming	Ozone Depletion	Toxics	Acid Rain	Occupational Health	Other	Energy Use	Durability/Maintenance	Recycling/Reuse/Disposal	Health	Other	
Ply-Based Boards																
Plywood	1	●	●	•	•		•	•	●				•	•	•	
Blockboard/Laminboard	0.9	●	•	•	•		•	•	●				•	•	•	
Particleboards																
Chipboard	0.5	●	•	●	●		●	●	●				•	●	●	
Oriented Strand Board (OSB)	-	●	•	●	●		●	●	●				•	●	●	
Fibreboards																
Softboard	0.3	•	•						•				•			
Mediumboard	-	•	•						•				•			
Hardboard	0.8	•	•						•				•			
Medium Density Fibreboard	-	●	•	●	●		●	●	●				•	●	●	
Cement-Bound Boards																
Wood-Cement Particleboard	-	●	•	●	●		•	•					●	●	•	Haz. Waste
Woodwood Cement Slabs	-	●	•	●	●		•	•					●	●	•	Haz. Waste
Decorative Laminates (Additional Impacts)																
Melamine	4	●		●			●	●	●				•		●	
Tropical Hardwood	4	●	●						•				•			

A Note on Durability

Durability of board products varies both between and within board types, depending on the resin used, the mode of manufacture, or in the case of ply and blockboard, the type of wood used. It is also dependent on the use to which the board is put. For example, as an internal insulating layer in dry conditions, softboard may last for tens of years, but it is inappropriate for situations exposed to weather, where it may last for only a few weeks. Similarly, it could be considered a waste of resources to use hardwood ply or a chipboard where a resin-free softwood fibreboard would suffice.

The Unit Price Multiplier

The price multiplier has been calculated for boards of equal width (12mm), using the cost of 12mm Russian Birch interior plywood as the basic price (£5.63m^2).[55] Only internal grade boards have been included; The multiplier for external grade ply and chipboards is around double that for interior.

8.4 Product Analysis

8.4.1 Synthetic Resins

The resin based boards listed in this report use mainly formaldehyde or diisocyanate resins and therefore have several impacts 'in common'. This introductory section outlines the impacts of these petrochemical based resins to save repetition in each material section.

Manufacture

Energy Use

The production of glues is highly energy consuming.[7]

Resource Use (non-bio)

The primary raw material for synthetic resin production is oil or gas, which are non-renewable resources.

Global Warming

Petrochemicals manufacture is a major source of NOx, CO_2, methane and other 'greenhouse' gasses.[50]

Toxics

The petrochemicals industry is responsible for over half of all emissions of toxics to the environment, releasing particulates, heavy metals, organic chemicals and scrubber effluents.[50]

Volatile organic compounds released during oil refining and further conversion into resins contribute to ozone formation in the lower atmosphere with consequent reduction in air quality. Emissions can be controlled, although evaporative loss from storage tanks and during transportation is difficult to reduce.[49]

Acid Rain

Nitrogen oxides and sulphur dioxide, involved in the formation of acid rain, are produced during refining and synthetic resin production.[49]

Other

The extraction, transportation and refining of oil for the production of products such as synthetic resins, can have enormous environmental effects,[50] such as the Sea Empress oil disaster at Milford Haven.

Use

Health

Formaldehyde resins are the most common bonding agent used in composite board manufacture, and have been found to yield measurable amounts of formaldehyde, particularly when the board has not been treated with an impermeable surface. It is not the amount of formaldehyde contained in the resin, but the amount of 'free formaldehyde' which is of importance. Free formaldehyde is formaldehyde which is not chemically bound within the resin and is available for off-gassing. Urea formaldehyde tends to contain the most free formaldehyde, while phenol and resorcinol formaldehydes tend to be more stable.[7] Formaldehyde is classed as an animal carcinogen and a probable human carcinogen.[20] Phenol formaldehyde has been linked to dermatitis, rashes and other skin diseases,[16,20] as well as respiratory complaints associated with exposure to component vapours.[20] Occupational exposure to synthetic glues based on Carbamine- and Phenol-formaldehyde resins have been linked to catarrhal respiratory disease and locomotive disorders.[17] (For more detail see the introduction, p.84).

8.4.2 Ply-based Boards

(a) Plywood

Manufacture

Energy Use

See 'synthetic resins' section above.

Energy is used in cutting the veneers, and in pressing, curing and drying during manufacture of the board.

Most plywoods are imported from the Americas or East and Far-East Asia,[1, 31] and so the high energy costs of transportation must also be taken into account, unless plywood from UK or European sources is specified.

Resource Use (non-bio)

See 'Synthetic Resins' Section

Resource Use (bio).

Plywood appears more efficient in its use of wood than sawn timber, although the efficiency varies depending on the species and process. Some species yield up to 90% usable veneer, some less than 30%.[7] Waste material is usually recycled into other types of board or burnt to provide energy.[7] However, unlike most board products, plywood does not utilize waste wood or sawmill wastes, requiring whole logs to produce large sheets of veneer.

Acid Rain & Global Warming

See 'synthetic resins' section.

Toxics

Plywood plants may emit large quantities of volatile organic compounds, largely as a result of their dryers.[29] (Also see 'synthetic resins' section).

Occupational Health

Dust can be a problem during manufacture.[7] Chronic long term exposure of workers to formaldehyde in plywood plants induces symptoms of chronic obstructive lung disease,[12] and plywood mill work is listed amongst the occupations with increased risk of birth defects in offspring.[13] Studies in Scandinavia suggest that fumes from the wood drying process may be carcinogenic.[14] There are also suggestions of risk of the disease Manganism, through manganese exposure during plywood manufacture.[46]

It is likely that these health hazards relate to the manufacture of other wood based board products where they include dust-creating cutting processes and the use of formaldehde-based glues. However, it appears that most of the research has concentrated on plywood.

Use

Health

Plywoods tend to use less glue then other board products and the nature of the curing process tends to leave less free formaldehyde.[7] The release of formaldehyde during use should therefore not be a problem if the plywood has been correctly manufactured to conform with BS 6100.[1,8] WBP plywoods are generally bonded with PF glue, which leaves little free formaldehyde, although problems have been reported with UF-bonded plywoods.[7] (See 'synthetic resins' section opposite).

Recycling/Reuse/Disposal

Reusable around five times before the surface deterioration prevents further use.[7] Burning may release harmful gases such as hydrogen cyanide from isocyanate resins.

Certification & Bogus Claims

Plywood and particleboard are currently not listed under the Soil Association timber certification scheme. Scandinavian composite boards and panels are covered by the 'White Swan' Joint Nordic Environmental Labelling Scheme.[32]

The Rainforest Alliance have certified the operations of plywood and veneer producers AMACOL Ltd, Portel, Para Brazil and Keweenaw Land Association Ltd, USA.[47] Buchners' 'EcoPanels' (TM) are hardwood veneers manufactured in San Francisco from forests certified by Scientific Certification Systems Inc. (SCS) or the Rainforest Alliance's Smartwood Program.[30] SCS also certify Collins Pine, Kane Hardwood, Pennsylvania, who produce hardwood veneers.[47]

Both Malaysia and Indonesia, two of the worlds biggest exporters of tropical plywood,[51, 52] now stamp all their plywood with the label 'Sustainable Timber'. In Indonesia, 1992 deforestation rates were 35km2 per day; Malaysia, according to its own government, will be a net timber importer by the end of the century due to depletion of rainforest resources.[31]

Incidentally, the UK Forests Forever campaign of the Timber Trade Federation has openly accepted a ''donation'' of £25,000 from Bob Hasan, Indonesia's largest plywood baron, who has built his multi-million pound fortune from unsustainable forestry and abuse of tribal landowners.[31]

(b) Blockboard/laminboards

Manufacture

<u>Energy Use</u>
See 'synthetic resins' section, p.88. Energy is also used in cutting the veneers, and in pressing, curing and drying during the manufacture of the board.

<u>Resource Use (non-bio)</u>
The raw materials for the production of synthetic resins are oil or gas, which are non-renewable resources.

<u>Resource Use (bio)</u>
The central core of these materials is generally softwood, and an increasing number of ply mills are introducing blockboard/laminboard manufacturing facilities to use residues from plywood manufacture for the central slab,[1] representing a good use of resources. The outer veneers for these boards are manufactured by the same process as for plywood (see above), and may come from any source, including tropical hardwoods.

<u>Acid Rain, Toxics & Global Warming</u>
See 'synthetic resins' section, p.88.

Use

<u>Health</u>
Almost invariably, blockboards and laminboards are manufactured using urea-formaldehyde adhesive,[1] which tends to contain more free formaldehyde than the other formaldehyde based resins.[7] However, as with plywood, escape of formaldehyde is likely to be limited by the structure of the board. (See 'synthetic resins' section, p.88 for more detail)

<u>Recycling/Reuse/Disposal</u>
Reusable around five times before the surface deterioration prevents further use. Burning may release harmful gases such as hydrogen cyanide from isocyanate resins.[7]

<u>Other</u>
See 'synthetic resins' section, p.88.

8.4.3 Particleboards

(a) Chipboard

Manufacture

<u>Energy Use</u>
See 'synthetic resins' section, p.88. Energy is also used in cutting the particles, mixing the particles with resin and pressing and drying the board.[32]

<u>Resource Depletion (non-bio)</u>
See 'synthetic resins' section, p.88.

<u>Resource Depletion (bio)</u>
Chipboard manufacture uses mainly softwood residues including forestry thinnings, planer and shavings and other joinery shop residues.[1] However, many chipboard manufacturers use whole logs.

A major source of wood for chipboard is plantation grown softwood.[7]

<u>Occupational Health</u>
See 'synthetic resins' section, p.88.

<u>Toxics</u>
Chipboard plants release large quantities of volatile organic compounds (VOCs) largely as a result of their dryers.[29] Also see 'synthetic resins' section.

Use

<u>Health</u>
The chemical composition of the resins is such that unstable components are mostly released during the first few days after manufacture.[8] A study in Denmark found that children living in houses with "much particleboard" have an increased risk of headaches, throat irritation and "need for daily antiasthmatic medication". Children under five also run an increased risk of developing wheezy bronchitis, eye and nose irritation. This risk was not only noted in new houses, but also in houses 8-14 years old.[43] However, "much particleboard" is defined as houses with walls, floors and ceilings made from particleboard. No risk was shown for homes with only minor particleboard content.[40]

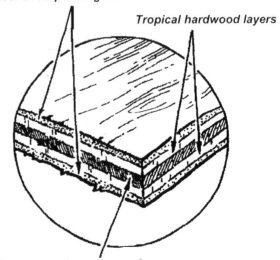

Veneer of corporate greenwash

Tropical hardwood layers

Compressed core layer of displaced wildlife, bonded with blood albumen from indigenous peoples

TYPICAL STRUCTURAL CROSS SECTION OF S.E. ASIAN PLYWOOD BOARD

Recycling/Reuse/Disposal
Reusable around five times before the surface deterioration prevents further use.[7] Burning may release harmful gases such as hydrogen cyanide from isocyanate resins.
Other
See 'synthetic resins' section, p.88.

(b) Oriented Strandboard (OSB)

Manufacture

Energy Use
See 'Synthetic Resins' section. Energy is also used in cutting the wood strands and forming the board using heat and pressure.
Resource Depletion (non-bio)
See 'Synthetic Resins' section, p.88.
Resource Depletion (bio)
OSB makes efficient use of wood, as none of the log is wasted and utilizes relatively uncontroversial species, such as aspen and pine.[28]
Occupational health
Workers in Oriented Strand Board plant, using aspen and balsam wood, with methylene diisocyanate (MDI) and phenol formaldehyde resins, have been found to suffer breathing difficulties, possibly due to the combination of wood dust, MDI and formaldehyde fumes.[15]
Toxics
OSB plants can emit large quantities of volatile organic compounds (VOCs) largely as a result of their dryers.[29] (Also see 'synthetic resins' section, p.88.)

In the USA, the large diameter softwood trees required to produce high quality plywood also provide habitat for the Northern Spotted Owl. In order to protect the owls, President Clinton has ruled that exploitation of the trees must cease. This has resulted in industry turning to Oriented Strand Board, which is produced from small diameter softwood and forestry thinnings. It has also boosted the UK Oriented Strand Board industry.
85% of Britain's Oriented Strand Board market is supplied by Norboard, an Inverness based company. The plant uses timber ''which would previously have been burned'' and small trees which quickly regenerate. Most of the plants energy needs are met by burning the bark stripped from the trees, and the timber is all taken from within a 40 mile radius.[18]

Use

Health
The off-gassing of formaldehyde from resins may cause health problems if allowed to gather in high concentrations[7] although OSB is generally bonded using phenol formaldehyde[40] which tends to be more stable than urea formaldehyde. (Also see chipboard section on the opposite page).
Durability
Increasing experiences of premature decay due to moisture are leading to questions about the suitability of OSB for use in weathering situations, particularly in humid climates. Manufacturers are currently exploring ways in which to improve its moisture resistance.[28]
Recycling/Reuse/Disposal
Reusable around five times before the surface deterioration prevents further use.[7] Burning may release harmful gases such as hydrogen cyanide from isocyanates.
Other
See 'synthetic resins' section, p.88.

8.4.4 Fibreboards

(a) Softboard
(b) Mediumboard
(c) Hardboard

Softboard may be impregnated with bitumen to improve moisture resistance, in which case the impacts of bitumen (not covered in this report) must be added to that of the board. Some dry-process hardboards may contain phenol formaldeyde resin to improve bonding strength.[40]

Manufacture

Energy Use
Energy is required to mechanically break down the wood into fibres and in the reconstitution process, which uses heat and pressure.[7]
Resource Depletion (bio)
Generally utilise softwoods, but hardwoods may be included.[7]
Occupational Health
Dust can be a problem in the manufacture of fibreboards, and can result in irritation of the skin, eyes and nose in areas with high dust levels.[7, 24]

Use

Recycling/Reuse/Disposal
The degraded condition of softboard on disposal may make re-use difficult.[7]

(d) Medium Density Fibreboard (MDF)

Manufacture

Energy Use

See 'synthetic resins' section, p.88. Energy is also used in cutting the particles, and in pressing, curing and drying during manufacture of the board.

Resource Depletion (non-bio)

See 'synthetic resins' section p.88. The resin content of MDF is greater than other particleboards at around 14%.[8]

Resource Depletion (bio)

Forest thinnings and undersized stock account for 85% of the material used in the manufacture of MDF.[7] $1m^3$ MDF uses approximately $1.4m^3$ raw wood, which is mainly softwood, although some hardwood may be utilised.[7]

Occupational Health

Dust can be a problem in the manufacture of fibreboards, and can result in irritation of the skin, eyes and nose in areas with high dust levels.[24] The effects with regard to resins are likely to be similar to other resin based particleboards.

Toxics

There may be some pollution of watercourses from effluents unless the plant is fitted with a closed water system.[7] (Also see 'synthetic resins' section p.88.)

Use

Health

Comments regarding chipboard and OSB are likely to apply equally to MDF, which contains a higher proportion of resin than other board products (see chipboard section and resins section relating to urea formaldehyde, the resin typically used in MDF.[40])

Recycling/Reuse/Disposal

MDF is difficult to use more than 3-5 times. It could be re-chipped but the high resin content may create problems.[7] Burning poses the same hazard as other resin bound boards.

Other

See 'synthetic resins' section, p.88.

An Ecology Group has been instituted by the Wood Panel Products Federation, to consider the environmental effects of the production and distribution of wood based board products, with special reference to chipboards, fibreboards , OSB and cement-bonded particle boards. Attention will be focused on life-cycle assessment of products, eco-labelling, and eco-auditing.[7]

8.4.6 Cement Bound Boards

(a) Wood-cement particleboard / (b) Wood-Wool Cement Slabs

Manufacture

Energy Use

Cement production is energy intensive, at 6.1 GJ/tonne for wet process kilns and 3.4 GJ/tonne for dry-process kilns,[45] due to firing at very high temperatures (~1400°C) using fossil fuels or waste (see Alert below).[1,2,5] Most transport of cement is by road, notable exceptions being the works at Dunbar, Hope and Weardale (all Blue Circle), which transport over 50% of their output by rail.[6] Energy is also required to cut and dry the wood particles/strands and in curing the board.

Resource Depletion (non-bio)

Raw materials for cement (chalk or Limestone) are fairly widespread, although suitable permitted reserves are running low in some areas - eg: the South East of England.[6]

Resource Depletion (bio)

The raw material for wood-wool boards is likely to be 'waste' from other wood processes.[19] and only softwoods are used - generally pine or spruce.[7] Wood-cement particleboard may contain some hardwood,[7] and it may not always be possible to tell if tropical timbers have been used.[19]

Global Warming

Production of cement is the only significant producer of CO_2 other than fossil fuel burning, responsible for 8-10% of global total (450million tonnes). CO_2 is given off during chemical reaction with calcium/magnesium carbonate materials (ie chalk or limestone).[3]

Toxics

Ordinary Portland Cement (OPC) contains heavy metals "of which a high proportion are lost to the atmosphere" on firing.[2] Organic hydrocarbons and carbon monoxide are also released, and fluorine compounds can also be present.[4]

Acid Rain

Sulphur Dioxide is produced in the cement kiln both as part of the chemical reaction of the raw material and as a product of burning fossil fuel. Normally, however, this is mostly reabsorbed into the cement by chemical combination and only a small amount escapes. Nitrogen oxides from the fuel are not absorbed.[4]

Other

The production processes for cement can cause a serious dust problem which is hard to control.[2] Extraction can also cause localised noise, vibration and visual impact.[5]

Use

Health Hazards

OPC dust released on sawing boards may contain free silicon dioxide crystals (the cause of silicosis), the trace element chromate (linked to stomach cancer and skin allergies) and the lime content may cause skin burns.[2]

Durability

Cement bonded particle board has good moisture resistance and is suitable for external use. It is also resistant to attack by wood boring insects in temporate climes.[7]

Wood-wool cement slabs are suitable for damp or high humidity conditions; air pockets within the slabs provide good breathing properties. They are resistant to rot, moulds, fungi and attack by insects and other animals.[44,9] Both types of board have increased dimensional stability over normal chipboard.[41]

Recycling/Reuse/Disposal

The recycling potential of wood-wool boards and wood-cement particleboard is extremely limited.[8]

Alert

All of the major OPC manufacturers are experimenting with substitute fuels derived from hazardous or toxic waste. This is a cause for concern because controls on emissions from cement kilns are much less stringent than for hazardous waste incinerators. For further information, see p.61 of Chapter 6.

8.4.5 Decorative Laminates

Manufacture

Energy Use

The production of oil based synthetic resins involves high energy processes, using oil or gas, which themselves have a high embodied energy.[39] Manufacture of the laminate and attachment to the board requires heat, to the order of 100°C.[38]

Resource Depletion (non-bio)

The raw materials for the resins (2,4,6-triamino-1,3,5-triazine, formaldehyde and phenol.[38]) are derived from crude oil, which is a non-renewable resource.

Resource Depletion (bio)

Melamine laminates use kraft or alpha-cellulose paper.[38] However, the softwood resources utilised in the production of the paper could be considered to be insignificant when compared to that required for the manufacture of the boards themselves.

Global Warming & Acid Rain

See 'synthetic resins' section, p.88.

Toxics

See 'synthetic resins' section, p.88.

Melamine also contains surfactants, plasticisers, release & anti-foam agents.[38]

Occupational Health

Manufacture of Melamine requires the heating of melamine powder with formaldehyde,[38] a probable human carcinogen.

Use

Durability

Melamine increases the durability of board products in terms of wear, moisture, heat and chemical resistance. They also increase the overall strength of a board.[38]

Recycling/Reuse/Disposal

Melamine increases the durability of a board and therefore increases the potential for re-use. Melamine is not biodegradable and, as a thermoset plastic, cannot be recycled through re-melting and reforming.

Other

See 'synthetic resins' section, p.88.

The Los Angeles Federal Reserve Bank is looking for new uses for over 7,000 tons of money it shreds each year. Gridcore systems proposes to used this durable, long fibre for producing fibreboard.[56]

Off-gassing of formaldehyde is most serious in warm locations, eg. near cookers or heaters, and where ventilation is restricted. It may be wise to avoid the use of formaldehyde emitting products in such areas.
The common spider plant, Chlorophytum comosum, removes formaldehyde from the air. It is a very easily maintained houseplant and reproduces more easily than almost any other plant. Keeping half a dozen in a room with newly fitted particleboard will diminish the effects of formaldehyde.[22]

8.5 Alternatives

(a) Tectan

A chipboard-like material made entirely from used beverage cartons!

The raw material for Tectan consists of process scrap from drink carton production, and used cartons collected from Germany's "Duales System Deutschland" (DSD) recycling system. The manufacturing process is similar to that of fibreboard production. Cartons are shredded, heated and cut into 5mm particles. The material is then spread into sheets, and bonded by heat and pressure, the polyethylene content acting as a glue. No additional materials are used other than the cartons themselves. If used uncovered, the boards have an interesting 'recycled' aesthetic, speckled with coloured particles and pieces of aluminium from the component packages (see picture) The board is composed of paper (75-80%), polyethylene (20%) and aluminium (5%). It has good insulation properties (W/m x K = 0.13) and water resistance, and can be sawn and screwed like conventional boards.[36]

The manufacturers claim that the material is superior to conventional building board due to its high noise absorption and resistance to moisture, and its thermoformable properties. Also, Tectan board can itself be recycled.[37]

While Tectan is an environmentalists dream in terms of its manufacturing impacts, tests by the AECB revealed problems with weight (it is extremely heavy) and fragile edges.[58] There are also problems of availability. The boards are manufactured at a small pilot plant in Germany by Entwicklungsgesellschaft fur Verbundmaterial Diez, mbH (EVD), a company owned by TetraPak.[36] At the moment, boards can only be purchased from the manufacturers and payed for in Deutschmarks.

Tectan is comparable in price to external quality hardwood ply or melamine faced chipboard, at around £8 - £9 (22DM) per m² for 6mm board - although prices may come down if the scale of production increases.

EVD Entwicklungsgesellschaft fur Verbundmaterial Diez, mbH
Industriestrasse 30
D-65582 Diez, Germany
Tel. 0 6432-1061
Fax. 0 6432-61826

Tetra Pak Ltd
1 Longwalk Road
Stockley Park
Uxbridge
Middlesex UB11 1DL
Tel. 01895 868000
Fax. 01895 868001

8.5.2 Straw

About 128 million tonnes of straw are produced in North America alone, and is mainly burned. If only 25% of this were used, it could produce 2.1 billion m^2 of 19mm particleboard - five times the current total U.S. production of particleboard.[27]

The appeal of straw is fourfold:

1. In areas of grain production, straw is inexpensive
2. Lumber supplies may be limited, prices fluctuate and the quality of lumber is dropping
3. The embodied energy of straw should be fairly low, as it is a secondary waste material from grain production
4. Using it as a building material would mean less straw would be burned in the fields, a practice which led to California's rice producers alone generating an estimated 51,000 tonnes of carbon monoxide each autumn, twice that produced by all of California's power stations.[27]

Environmental Building News reports that there are at least ten companies either currently building or planning to build plants in North America to manufacture compressed straw building panels.[27] The applications of strawboard range from interior partitions to particleboard. In many areas, straw is tilled back into the soil, which while providing few nutrients, helps aerate the soil and adds organic matter. There are concerns that without careful management, removing the straw for other uses would eventually have significant detrimental effects on soil structure and crop yields. However, there is also evidence that too much straw in the soil can upset the balance between soil bacteria and fungi, reducing soil fertility.[27]

(a) Thick Panel Products - STRAMIT

The process for producing "Compressed Agricultural Fiber" (CAF) panels was invented in Sweden in 1935, and was commercially developed in Britain in the late 1940s, under the name Stramit. The patents on the production technology have since expired, leading to an expansion of Stramit manufacture around the world.[27]

The process involves compression of straw at around 200°C, causing the straw fibres to bind together without any adhesives - a similar process to the production of fibreboard.

Stramit boards range in thickness from 50mm to 100mm and are faced with heavy weight kraft paper - the fixing of which requires the use of adhesives.[27]

Stramit is used primarily for interior applications such as partition walls, although some companies are currently developing structural insulating panels which can be used in the exterior envelope of buildings.

The material is low density, so the environmental and financial costs of shipping are high. It is estimated that the raw material could be shipped 18-20 miles to the production plant before shipping costs become a major economic obstacle.[27]

(b) Wheat-Straw Particleboard

Thin panels between 3mm and 13mm thickness, which are made from chopped wheat straw, mixed with a resin and pressed into panels. Many manufacturers (eg: Naturall) use non-formaldehyde, non-offgassing resins - usually 'MDI' resin,[28] although these isocyanate based resins have other environmental impacts, mainly during manufacture (see synthetic resins section, p.88).

Studies in North America suggest that wheat straw particleboard may be "superior to wood-based particleboard in moisture resistance and structural properties", and the manufacturers of Naturall Fibre Boards claim that their product is "close to the strength of plywood".[28] Small scale independent tests found that samples did not delaminate at all after days soaked in water, and that screws do not pull out as easily as they do with conventional particleboard.[28] While wheat straw particleboard still requires the use of resins, and energy is required to heat and press the board, it is potentially more environmentally benign than its wood based counterparts as it uses what is essentially a waste product as its primary raw material. Prices are also extremely competitive in relation to standard particleboard.

8.5.3 Trex Lumber

An alternative decking material made from recycled plastic bags, industrial stretch film, sawdust from furniture factories and used wood pallets. Trex Lumber can be drilled, sawed, painted and sanded like normal wood and is also suitable for underwater use. However, it is not as strong as wood.[57] Unfortunately, we have been unable to obtain an address for the manufactureres.

8.6 Specialist Suppliers

(a) Particleboards

'MINERIT' Composite Board

Cambrit, Suite C, Hamard House, Cardiff Road, Barry, South Glamorgan, CF63 2BE (tel: 01446 742095 fax: 01446 721041)

General purpose building board composing of cellulose fibres, portland cement and mineral fillers. Suitable for interior or semi-exposed situations.

CSC Caberboard Ltd, Station Road, Cowie, Sterling FK7 7BQ. (tel: 01786 812921 fax: 01786 815622)

Manufactured in Scotland from mainly scots pine grown from within a 40 mile radius of the mill. Composition is 96% softwood, 1.5% paraffin slack wax and 2.5% phenol formaldehyde resin (a lower formaldehyde content than even 'low formaldehyde' chipboard). Any timber waste from the process is used to heat the mill or goes to horticultural uses to reduce demand for peat compost.

Also produce oil tempered hardboard in varying thicknesses; Eco-attributes include zero-formaldehyde, natural content and low emissions.

Hornitex UK Sales Ltd, 2nd Floor, The Graftons, Stamford New Road, Cheshire, WA14 1DQ. (tel: 0161 941 3036 fax: 0161 928 9414)

Hornitex boards are manufactured to E1 (low formaldehyde content) standards. Waste timber from production is used to produce energy to run the plants. Hornitex also filter all exhaust gases from their production processes to remove dust and fumes. 20% of the companies investments are spent on environmental protection. The timber supply for all production comes from German forestry comission controlled woodstock and commercial quality woodwaste from other industries.

(b) Fibreboard

Masonite CP, West Wing, Jason House, Kerryhill, Horsforth, Leeds, West Yorkshire, LS18 4JR. (tel: 01132 587689, fax: 01132 590015)

Produce oil tempered hardboard in varying thicknesses; Eco-attributes include zero-formaldehyde, natural content and low emissions.

Also, low formaldehyde, low toxicity **Fibreboard** manufactured from extra-long-fibre wood material, pressed to high density (940kg/m3) to give high strength. Certified to ISO 9001

'Colourboard' noticeboard/pinboard

Celotex Ltd, Warwick House, 27-31 St Marys Road, Ealing, London W5 5PR. (tel: 0181 5790811 fax: 0181 579 0106)

Colourboard is recycled from waste newsprint and dyed with vegetable based dyes, making an environmentally friendly product. The board has class '0' fire performance. Also produce 'Mediumboard' pinboard from recycled newsprint, which can be used as a pinboard or as acoustic screening and wall linings due to its 'excellent thermal and acoustic insulation properties'.

'Bitvent 15' impregnated Fibreboard

Hunton Fibre UK Ltd, Market Chambers, 22a High Street, Irthlingborough, Northamptonshire, NN9 5TN (tel: 01933 651811 fax: 01933 652747)

Bitvent 15 is a sheathing panel designed for use in the "Breathing Wall", a system whereby moisture is able to penetrate the wall structure in a controlled manner. The board has been developed to meet the standards of the UK Timber Frame industry. It is durable, vapour permeable is easily cut and fixed, and has thermal conductivity of 0.055.

The boards are produced in Norway from wood chips and selected wood waste, using predominantly timber which would otherwise be burnt or wasted. Energy to produce the board is generated by Norways hydro-electric generating systems.

Guide Price: £8.00 per sheet

'PANELITE BITVENT 15' Bitumen impregnated fibreboard

Panel Agency Ltd, 17 Upper Street North, New Ash Green, Longfield, Kent, DA3 8JR (tel: 01474 872578 fax: 01474 872426)

Timber frame sheathing panel produced by 'Hunton Fiber', specially designed for use in the 'Breathing Wall' system. Hunton also produce a wide range of other fibreboard products including tongued and grooved.

'Fillaboard' bitumen impregnated fibreboard.

Filcrete Ltd, Grindell street, Heon Road, Hull, HU9 1RT (tel: 01482 223405 fax: 01482 327957)

A range of 'ecologically friendly' boards with low embodied energy, manufactured from timber by-products of other wood processing activities, with no adhesives. Certified to meet BS, DOT, DOE and BAA specifications and has thermal conductivity of 0.055. Water used during manufacture is recycled.

'Panelite Bitvent 15' impregnated insulating board

Falcon Panel Products Ltd, Unit C1A, The Dolphin Estate, Windmill Rd, West Sudbury on Thames, Middlesex, TW16 7HE (tel: 01932 770123 fax: 01932 783700)

Manufactured for use in 'Breathing Wall' systems. Certified to BBA standards.

(c) Flax & Straw Boards

'Canberra' flax based partition boards, 40-46mm thick

Stramit Industries Ltd, Yaxley, Eye, Suffolk IP23 8BW (tel: 01379 783465 fax: 01379 783659)

Contact: Sara Slade, Sales & Marketing

Unfaced or hardboard faced hardwearing solid partition medium for commercial and industrial uses where a robust partition is required. Certified to ISO 9001 and BS 5750, and has thermal conductivity of 0.101.

Guide Price £12.50m2

'EASIWALL' 58mm compressed straw boards

Stramit Industries Ltd (see above)

A range of strawboard fabricated using a petented process of heat and pressure with no adhesives. Conforms to BS 4046 and suitable for heights of up to 2.4 metres withg out additional support. Panels are 58mm thick and 1200mm wide, and have thermal conductivity of 0.101.

(d) Cement Bound Board

'Heraklith-M & BM' Woodwool Boards

Heraklith UK Ltd, Broadway House, 21 Broadway, Maidenhead, Kent SL6 1NJ (tel: 01628 784330 fax: 01628 74788)

Contact: James Muir

Manufactured from softwood (pine, spruce and poplar) and magnesite (a portland cement alternative) to give excellent sound and thermal insulation properties. Thermal conductivity = 0.073.

Woodcemair Woodwool board

Torvale Building Products, Pembridge, Leominster, Hereford, Worcester, HR6 9LA (tel: 01544 388262 fax: 01544 388568)

Contact: J.K. Richards, Technical Director

Wood wool cement slabs, 'composed entirely of natural non-toxic materials'. Embodied energy: 900 kWh/m3; Thermal Conductivity: 0.073. Certified to BS 1105 1981.

8.7 References

1. Construction Materials Reference Book (D.K. Doran, Ed.) Butterworth - Heinmann Ltd, Oxford 1992.
2. Greener Building Products and Services Directory (K. Hall & P. Warm) Association for Environment Concious Building, Coaley.
3. The World Environment 1972-1992 -two decades of challenge (M.K. Tolba et al, Eds) UNEP/Chapman & Hall Ltd, London, 1992
4. Technical Note on Best Available Technologies Not Entailing Excessive Cost for the Manufacture of Cement (EUR 13005 EN), Commission of the European Communities, Brussels 1990
5. Eco-renovation - the ecological home improvement guide (E. Harland) Green Books Ltd, Darlington 1993
6. Minerals Planning Guidance: Provision of Raw Materials for the Cement Industry (Dept of the Environment & the Welsh Office) 1991
7. Environmental Impact of Building Materials. Vol. E: Timber and Timber Products (J. Newton & R. Venables) CIRIA June 1995.
8. The Green Construction Handbook (Ove Arup & Partners/JT Design Build) Cedar Press Ltd 1993.
9. Materials, 5th Edn. - Mitchells Building Series (A. Everett & C.M.H. Barritt). Longman Scientific & Technical, Essex. 1994
10. At the Mercy of the Timber Barons (D. Young) The Times, 16th March 1995.
11. Where have all the Ceibas Gone? (A.H. Gentry & K. Vasquez) Forest Ecology and management 23 (8) p.73-76 1988.
12. Respiratory health of Plywood Workers Occupationally Exposed to Formaldehyde. (T. Malake & A.M. Kodama) Arch. Environmental Health 45 (5) p.288-294 1990.
13. Paternal Occupation and Congenital Anomolies in Offspring. (A.F. Olsham, K. Teschke & P.A. Baird) Americal Journal of Industrial Medecine, 20 (4) 447-475. 1991
14. Chromosome abberations in Peripheral Lymphocytes of Workers Employed in the Plywood Industry (P. Kuritto et al) Scandinavian Journal Work. Env. Health 19 (2) p.132-134. 1993.
15. Pulmonary Effects of Simultaneous Exposures to MDI, Formaldehyde and Wood Dust on Workers in and Oriented Strand Board Plant. (F.A. Herbert, P.A. Hessel, L.S. Melenka et al) Journal of Occupational Environmental Medicine 37 (4) p.461-5. April 1995.
16. Effect of the Working Environment on Occupational Skin Disease Development in Workers Processing Rockwool. (M. Kiec-Swiercsynska & W. Szymczk) International Journal Occup. Med. Env. Health 8 (1) p.17-22. 1995.
17. Occupational Hygiene in the Plywood Industry. (M.E. Ickovskaia) Med. Tr. Prom. Ekol. 11-12 p.20-22.
18. The American Owl that Saved Britain a Fortune. (D. Young) The Times, 16 March 1995
19. Green Building Digest, Issue 2, Feb 1995.
20. Buildings and Health - the Rosehaugh Guide to the Design, Construction, Use and Management of Buildings. (S. Curwell, C. March & R. Venables) RIBA Publications, London, 1990
21. C for Chemicals (M. Birkin & B. Price). Merlin Press Ltd, London.. 1989
22. Eco-Renovation (E. Harland). Green Books, Devon. 1993
23. Dictionary of Building (J.H. Maclean & J.S. Scott) Penguin, 1993
24. Occupation Irritant Contact Dermatitis from Fibreboard Containing Urea Formaldehyde Resin (P.T. Vale & R. Ryecroft). Contact Dermatitis 19 (1). 1988

25. Proving Chemically Induced Asthma Symptoms - Reactive Airways Dysfunction Syndrome, a New Medical Development. (R. Alexander) http/www.seamless.com/talt/txt/asthma/html. 1996

26. Reactive Airways Dysfunction Syndrome (RADS). (S.M. Brooks,
M.A. Weiss & I.L. Bernstein). Chest 88 (3) 376-384. 1985

27. Environmental Building News 4 (3) May/June 1995

28. Environmental Building News 4 (6) Nov/Dec 1995

29. Environmental Building News 4 (4) July/August 1995

30. Environmental Building News 4 (2) March/April 1995

31. Building for a Future 2 (1) p.17 Spring 1992

32. Environmental Impact of Materials Volume A - Summary, CIRIA Special Publication 116 1995.

33. Caberboard Environmental Policy Statement EP 101 Rev 3 (Caberboard, South Molton) February 1996.

34. Respiratory Sensitisers - A Guide to Employers (COSHH). Health & Safety Executive (HSE) 1992

35. BREEAM - An Environmental Assessment for New Homes. Version 3/91 (J.J. Prior, G.J. Raw & J.L. Charlesworth) Building Research Establishment (BRE) 1991

36. TECTAN by Tetra Pak - The Clever Solution (Tetra Pak Ltd) (undated)

37. TECTAN - Beverage Carton Recycling In Action (Tetra Pak Ltd) (undated)

38. Plastics: Surface and Finish. (W. Gordon Simpson, Ed.) Royal Society of Chemistry, Cambridge 1993

39. Achieving Zero Dioxin - an emergency stratagy for dioxin elimination. Greenpeace International, London. 1994

40. Hardwood Plywood & Veneer Association: Terms of the Trade. http://www.access.digex.net/~hpval/terms.html (HPVA) (1996)

41. Timber - Its Structure, Properties and Utilisation (H.E. Desch & J.M. Dinwoodie) Macmillan Press Ltd 1981

42. Occupational Exposure Limits EH40/94. HSE. 1994

43. Is Particleboard in the Home Detrimental to Health? (P. Daugbjerg) Environmental Research 48 p154-163. 1989

44. The Building Design Easibrief. (H. Haverstock) Morgan-Grampian Press. Jan 1993

45. Environmental Impact of Building and Construction Materials Volume B - Mineral Products (Wimpey Environmental Ltd). CIRIA, June 1995.

46. Manganese Exposure in the Manufacture of Plywood - An Unsuspected Health Hazard. (E.J. Esswein) Applied Occupational Env. Hygene 9 (11) p745-751. 1994

47. Green Building Digest Issue 4 - Timber (p13). July 1995

48. Occupational Exposure Limits EH40/94 (HSE) 1994

49. Environmental Impact of Building and Construction Materials - Volume F: Paints & Coatings, Adhesives and Sealants (R. Bradley, A. Griffiths & M. Levitt) CIRIA, June 1995

50. The World Environment 1972-1992. Two Decades of Challenge. (M.K. Tolba & O.A. El-Kholy (Eds.)) Chapman & Hall, London. 1992

51. U.S. Demand for Indonesian Plywood. (I.B.P. Parthama & J.R. Vincent). Bulletin of Indonesian Economic Srtudies 28 (1) 101-112. 1992.

52. A Review of Tropical Hardwood Consumption (J.D. Brazier) BRE Publications 1992.

53. Timber & Joinery - Keynote Market Report. (E. Caines (Ed.)) 1996.

54. Simply Build Green - A Technical Guide to the Ecological Houses at the Findhorn Foundation. (J. Talbott). Findhorn Press, Forres, 1995.

55. Spons Architects & Builders Price Book. (Davis, Langton & Everest, (Eds)). E & FN Spon, 1995.

56. New York Times p.1, May 22 1994

57. Journal of Light Construction p.40 August 1994

58. Building for a Future, Vol.2 No.6. Summer 1996.

Timber Preservatives 9

9.1 Scope of this Chapter

This chapter examines the environmental impacts of the most common wood preserving chemicals available on the market. Products covered include creosote, copper, chrome, arsenic, zinc boron and fluorine compounds, pentachlorophenol, dieldrin, lindane, tributyl tin oxide and permethrin. Due to the extremely large number of patented formulations, we have looked at preservatives in terms of the main chemicals used, the solvent (water, organic or other) and the mode of application. These can easily be related to specific products which, under the Control of Pesticides Act 1986, must have an ingredients list printed on the container.

9.2 Introduction

9.2.1 What are preservatives?

Wood preservatives comprise a mixture of solvent (organic or water) and active ingredients such as pentachlorophenol (PCP), arsenic, chrome and tributyl tin oxide (TBTO).[7,8] Preservatives are also available as sticks, pastes and 'smokes'.

Timber preservatives are, by their very nature, highly persistent and ALWAYS toxic - some extremely so.[34,46] Most preservatives have been cited as causes of ill health, based on well documented toxicological effects due to occupational exposure of wood treatment operatives.[7]

Exposure to the users of treated buildings is usually through the inhalation of dust particles which have the compound attached to them. Woodworkers will also be exposed to sawdust from preservative treated wood, and skin contact through carrying treated timber.[7,8] The groups most at risk are workers involved with the pretreatment and remedial treatment of timber.

As well as being irritants and nerve poisons, some of these compounds target organs such as the liver and can accumulate there from a number of sources, such as non-organic agricultural produce, as well as from building sources.[7]

Such treatments may not always be necessary at all, and there may also be a non-toxic or less toxic treatment technique available.[46] (see Alternatives section, page 114).

9.2.2 Types of Preservative

Tar-oil preservatives

(BS 144:1973 & BS 3051:1972)

Creosote, produced by distillation of coal tar, is the most common preservative in this category and is highly toxic to fungi, insects and marine borers. Heavy-oil creosote is used for high hazard situations such as telegraph poles, marine and freshwater pilings and railway sleepers. Medium and light oil creosote are typically used on fencing and farm buildings. All forms of creosote have a characteristic smell which remains for long periods after application, and treated timber cannot usually be painted.[2]

Water-borne Preservatives

Solutions of a single salt or mixtures of salts in water, of which there are two categories - fixed salt treatments, and water-soluble treatments.

The salts in fixed salt treatments react with the timber to become insoluble. They are usually applied by high pressure process, after which the wood is usually re-dried, a process which may result in dimensional changes in the timber.

The majority of water soluble treatments are boron compounds in water. Boron compounds are fungicidal

and insecticidal, but can be leached from the timber, so their use is restricted to dry areas where leaching is not expected, such as internal structural timbers.[2]

Mixtures containing combinations of copper, chrome, arsenic zinc or fluoride are also popular.

Organic/Solvent Borne Preservatives

(BS 5707)

Consist of fungicides and/or insecticides dissolved in an organic solvent. Most are highly resistant to leaching, and are suitable for low to medium hazard situations.[2]

Common organic preservatives are pentachlorophenol (PCP), organic zinc copper and tin compounds, lindane (y-HCH) and permethrin.

The organic solvents used as carriers may be volatile, or relatively non-volatile petroleum fractions.[3] Most are applied by low-pressure processes or by immersion, and cause no dimensional movement of the timber. After evaporation of the solvent, the timber is no more flammable than untreated wood.

Organic preservatives may be modified to include water repellents, tints for recognition, or other additives. Most are compatible with paints and glues.[2]

Pastes

Emulsions applied as thick pastes to the surface of timber, where deep penetration or precisely placed concentrations of preservative are required.[34]

Insecticidal Smokes

Used for the control of insects such as deathwatch beetle, where there is a need to kill emerging adults. Normally carried out as an annual treatment.[34]

Solid Plugs

Fused rods of soluble boron compounds, inserted into pre-drilled holes, which are dissolved by moisture and diffuse through the wood. These have a fungicidal action and are popular for use in window frames and similar products.[34]

Surface Coatings

Organic solvent based protective paints, which protect the wood by preventing moisture ingress. These sometimes contain fungicides.[34]

9.2.3 The Main Issues

Many preservatives, particularly those developed some time ago, are highly hazardous to health and 'must be handled and used with the utmost care'.[7]

There is potential exposure to wood preservatives at every stage of construction from site preparation to occupation.[7] Maintenance work, where preservatives are applied in-situ, is likely to present the greatest hazard.

Dry rot, wet rot and many wood boring insects will only occur in damp timber. Solve your damp problem and you have gone a long way to solving your pest problem.[46]

The development of a specialist wood preserving industry over the last 40 years has had the effect of allowing professionals such as surveyors, architects and general builders, to ignore the problem of timber decay by providing an instant spray-on 'solution'.[61]

Active Ingredients	Health	Environment	Fire	Disposal	Other	Banned (in UK)
Creosote	●	●	·		●	
Boron	·					
Arsenic	●	●		●	●	
Chromium Salts	●	●		●		
Copper Slats	·			●		
Fluorides	●					
Dieldrin	●	·	●	●	●	X
PCP	●	●	●	●	●	
Lindane	●	●	●	●	●	
Dichofluanid	·	·	●	●	●	
Tributyl Tin Oxide	●	●				
Permethrin	·	●	●			
Cu & Zn Naphthanates	●					
Acypectas Zinc	●					
Carriers:						
Water Based					·	
Solvent Based	●	●	●		●	

Table title: **Timber Preservatives**

9.3 Best Buys

The environmental 'Best Buy' is well detailed, properly seasoned wood with a protective finish, in buildings designed with adequate ventilation and avoidance of moisture sources. In such a situation, the use of preservative chemicals is usually unnecessary.

Overall preservation of wood is hardly ever necessary, but timbers in 'high risk' situations such as window sills may require localised treatment.

If preservatives are required, the best option is boron compounds, which appear to have a low toxicity to humans and the environment, followed by water based zinc/copper/fluoride compounds.

Solvent borne organic preservatives should avoided, as both the solvent and preservative tend to be volatile, leading to an inhalation hazard. Lindane, pentachlorophenol and tributyl tin oxide are particularly toxic and should be avoided.

Formulations containing chrome and/or arsenic (eg: Copper chrome arsenate) and permethrin are slightly less hazardous, the former due to its stability in the wood once dry, and the latter due to a slightly lower toxicity than the alternative synthetic preservatives.

9.3.1 When to Treat:

Pretreatment appears to be the least hazardous form of application, as it is carried out at specialist plant. On site treatment should be avoided, with the exception of localised treatment with low hazard preservatives such as boron rods.

Best Buys - At a Glance

First Choice: No preservative

Second Choice: Water based boron or zinc, copper, and/or fluoride compounds

Try to avoid: Chrome/Arsenic compounds, permethrin

Avoid: Lindane, pentachlorophenol, tributyl tin oxide, Creosote and (dieldrin)

Note: Many of the active ingredients listed are only available in combination with other active ingredients - eg Copper chrome arsenate. However, due to the huge range of formulations on the market, the impacts of each ingredient have been listed separately.

NB: Wood stains often contain preservatives, for which the same 'best buys' apply.

9.3.2 Application Methods:

In order of decreasing risk:

Smokes
Spraying
Dipping timber or joinery in tanks
Pressure impregnation
Brushing
Spreading/mastic-gunning pastes/mayonnaises
Drilling and injecting jellies
Drilling and inserting rods

(Source:22)

Water based paste or solutions are recommended as the safest method of application after borax/boron rods.[7] Organic solvents can be a "safe and specific" method of application[7] although many solvents are potential hazards in themselves, and tend to be volatile.[7]

Pressure treatment in pretreatment plants is recommended by the AECB as the most controlled way of using preservatives. The timber is impregnated properly and any waste is collected and re-used.[29]

Dipping should be relatively safe, although bad management of the job can lead to high exposure. Studies of US timber treatment workers confirm that dipping operations can lead to some of the highest levels of PCP absorption.[22]

If these methods are not suitable, spraying is probably the next best option but only if it is carried out with due regard to the potential hazards.[7] Spraying creates a high concentration of mist or vapour ideal for both inhalation and absorption through the skin.[22] It is also less precise, leading to contamination of non-target areas.

Smoke bomb applications are indiscriminate, uncontrolled and best avoided, and their efficiency has been questioned.[7]

On site use should be kept to an absolute minimum as it is very difficult to control and can therefore be extremely hazardous. Unpredictable weather conditions can result in seepage into watercourses, and lack of care resulting in spillage can cause problems of ground and groundwater contamination.[29]

EFFECTIVENESS

In 23 year exposure tests, using L-joints to simulate window and door external joinary, BRE ranked the performance of preservatives in the following order;

Creosote
PCP
Copper & Zinc Naphthenate
1% TnBTO.

The addition of water repellents did not cause any consistent and long lasting improvement in long term performance.[32]
Untreated timbers were ranked in the following order (decreasing durability);

Hemlock
Beech
Whitewood
Pine Sapwood

This order was the same for timbers treated with the same preservative.[32]

9.4 Impact Analysis

9.4.1 Pretreatment, Remedial Treatment, Professional or DIY?

(a) Pretreatment

BRE recommend that where timber treatment is unavoidable, specification should favour pre-treatment, as this is carried out in specialized industrial plants by trained specialists,[1,4] subject to ''rigorous health and safety checks''[1] by HSE inspectors and emissions controlled by integrated pollution control (IPC) guidelines,[36] whereas on-site treatment is often carried out by non-specialist personnel.[4] Pre-treatment plants are also more likely to have good trade union organisation.[22]

Pre-treatment in a specialised plant is also claimed to reduce the potential for solvent emissions into the building after completion,[4] and to allow more efficient and effective use of the preservative chemicals.[1]

On the other hand, pretreatment presents additional risks due to the huge amounts of chemicals present on site and the high pressures used to drive preservatives into wood. Although under normal operating conditions the airborne pollution risk is less than for remedial work, handling wet timber can lead to chronic or acute poisoning.[22]

(b) Remedial Treatment

This presents the greatest risk to operatives and building users. Before remedial treatment, curtains, carpets and soft furnishings should be removed. Care must be taken not to contaminate water supply or food.[7]

The treated area should be inspected to ensure that there are no exposed areas with unwanted contamination. The area should be left ventilated but unoccupied for at least 24 hours to allow airborne pesticide levels to drop.[7]

Some of the carriers used in wood preservatives are volatile and have an odour of their own - often mistaken for the preservative by the occupier, leading to a loss of confidence in the contractor. It may be best to delay reoccupation until the area is odour free, not least because some carriers are irritants in themselves.[7]

Timber treatment contractors have no legal requirement to undergo any training or hold any certificates.[46] It is therefore possible for anyone to set themselves up as ''specialists'' in the field of remedial treatment, without training or prior experience.[22] It is worth checking if a contractor is a member of the British Wood Preserving and Damp Proofing Association (BWPDPA), which tries to maintain high professional standards in the industry, and eliminate 'cowboys'.[46]

Health & Safety

It is only jobs involving building work and lasting over 6 weeks that have to be notified to the HSE. Spraying of preservatives is not notifiable, even under the Control of Pesticides regulations. There is therefore no mechanism to inform enforcers that a home is about to be treated, and the prospects for law enforcement are therefore poor for small scale remedial work.[22]

Once a property or workplace has been treated, any problems should be reported to the Environmental Health department of the local council.[22]

(c) DIY

For householders, the greatest risk is from DIY work without full protection.[8] PEGS (see page 111) claim that the majority of callers with preservative-related complaints were exposed during DIY work.

9.4.2 Preservative Carriers/ Solvents

(a) Water-Borne Preservatives

Health

Waterborne preservatives tend to be less volatile than solvent borne, so pose less of a risk through inhalation of vapours.

Other

Waterborne preservatives are odourless and non-combustible when dry.[3]

They tend to cause timber to swell during treatment, which can lead to warping.[30]

(b) Organic Solvent- Borne Preservatives

In some wood preservative fluids, the solvent may make up over 90% of the formulation.[22]

Health

Pungent solvents in which many preservatives are formulated have had complaints such as sore eyes, chest symptoms and headaches attributed to them.[7]

Organic solvents are known to affect the brain and nervous system, causing narcosis (drunkenness), memory loss, slowing of thought, slow reflexes, loss of feeling or movement in extremities, and tremors.[22]

Chronic exposure or high dose acute exposure to solvents can also damage the liver, kidneys, digestive system, eyes, respiratory system and skin, causing irritation, allergy and possible long term damage including dermatitis and pneumonitis.[22, 56]

Solvents used in preservatives are blends of many compounds including paraffins, which have the potential to cause cancer. Serious ailments such as Prader-willi syndrome and childhood cancer are far more common in the offspring of solvent or pesticide exposed workers than for the general population.[22]

Solvents may also have synergistic effects with other

Local Authorities

Local authorities are the largest single group of pesticide users in the UK outside agriculture. Council applied pesticides - including wood preservatives - are applied in and around homes, schools and residential institutions used by some of the most vulnerable members of society, and those potentially most at risk from the effects of pesticides, such as children, the elderly and the sick. Most council officials assume that their use of pesticides is insignificant, and many fail to make the link between their pesticides/preservative use and the Agenda 21 policies which they are currently developing.[47]

toxins.[56]

It has been suggested that the use of solvent based preservatives should be restricted to areas where the fumes cannot enter the living space, such as underfloor areas.[7]

Fire & Explosion

Organic solvents are highly flammable, although after evaporation of the solvent, treated timber is no more flammable than untreated wood.[2]

Environment

Organic solvents are derived from petrochemicals refining,[3] for which the raw material is oil, a non-renewable resource. The production of, and offgassing of volatiles from organic solvents make a significant contribution to the greenhouse effect.[56]

Other

Many solvent based preservatives are subject to loss by evaporation, which reduces their protective properties over time.[30] Some have an odour which may taint food.[3] Organic solvent based preservatives give wood a weathered, silver-grey appearance and are not suitable for preserving a natural finish.[31]

9.4.3 Active Ingredients

(a) Coal Tar Creosote

The most common DIY preservative,[22] incorporating the active ingredient and solvent in one material.[33]

Health

The chemical composition will vary depending on the stage in the distillation process that the oil is drawn off.

Reducing Preservative Offgassing

Polyurethane varnishes painted onto treated wood can reduce PCP emissions by 90-95%, and latex paint reduces it by 84%,[27,28] although it must be remembered that polyurethane has its own impacts (see chapters 5 & 15). Paint, however, increases offgassing, as the preservative dissolves into the paint.[22]

Over-painting can be effective where only a limited amount of preservative has been used.[22]

Consequently, the effectiveness of the material and its attendant health hazard can vary enormously.[33]

The odour of creosote makes it unsuitable for internal use, particularly near food.[3]

Creosote has been shown to have synergistic effects with other chemicals, such as 2,6-dinitrotoluene (DNT), with which it forms carcinogenic and mutagenic metabolites in rats.[19]

Creosote is reported to cause skin and eye irritation,[22, 29, 33] headaches and nausea, skin cancers[33] and permanent damage to the cornea.[22, 29, 34] Irritation is made worse by sunlight.[29] Creosote vapours consist up to 80% polynuclear aromatic hydrocarbons (PAHs), many of which are carcinogenic and genotoxic,[35] and inhalation has been linked to acute bronchitis and cancer.[22, 34] The creosote industry is thought to be one of the major sources of PAH in the UK, accounting for up to 25,000 tonnes per year.[38] The release of both solvent and toxin will occur for a considerable time after application.[33]

The risk of high exposure to DIYers is great, and creosote has been banned for some time in the US for all but professional use.[34]

Fire & Explosion

Creosote is not readily flammable, and any increased fire risk decreases within a few months as volatiles are lost.[3]

Environment

The coal products industry, including creosote timber treatment plants, have left a legacy of land contamination - estimated to contain between 8,000 and 80,000 tonnes of PAH.[38]

Creosote is injurious to some forms of plant life.[3]

Disposal

Creosote is biodegradable.[53]

Other

Creosote treated timber cannot be overpainted, although creosote itself has a limited decorative effect.[31] It is not as stable as other preservatives and can bleed out of timber, particularly in damp or wet situations.[53]

Creosote is generally not corrosive to metals.[3]

(b) Boron Compounds

Available in numerous formulations for use as insecticides, fungicides herbicides and disinfectants.[46]

Health

Research suggests that these are probably the least harmful of chemical timber treatments.[46, 65] The London Hazards Centre suggest that these can be used safely in pretreatment, and as solid rods in remedial work.[22] Nevertheless, boron compounds should not be considered as completely safe. At very high concentrations, boron compounds are neurotoxins and attack the liver, kidneys and lungs, and fatalities have occurred through swallowing them.[46] Acute symptoms include skin rash and a fall in body temperature. Chronic symptoms include damage to the nervous system, changes in body chemicals and allergic reactions. Boron compounds can affect reproduction and can be foetotoxic.[46,56]

Other

Boron compounds are water soluble and tend to collect in the dampest areas of the wood, which is advantageous as these tend to be the areas of optimum fungi and insect habitat.[65]

There is no discoloration or other change to the surface of the wood due to boron treatment.[65]

(c) Arsenic Based Chemicals (CCA)

Arsenic is an insecticide and fungicide used in pretreatment only,[22,29] usually in conjunction with copper and chrome (CCA),[22] (eg: the Tanlith process.)[29]

Health

Arsenic is classed as a 'deadly poison'[22, 34], the lethal dose for adults being around 500mg.[22] Arsenic causes damage to skin and peripheral nerves resulting in loss of movement

Legislation

Wood preservatives are covered under the Control of Pesticides Regulations 1986. This prevents the advertising, sale, supply, storage and use of preservatives unless they have been granted official 'approval' after the submission of safety data, and they meet the general requirements set out in formal 'Consents' published under the regulations.[7] Legislation on preservatives in the UK rest on the control of active ingredients, rather than on the control of products,[43] and preservatives must meet strict labelling requirements, including a complete list of ingredients.[7] Some timber treatment chemicals are restricted to use by 'professionals', although since 'professionals' need no official certification, effectively anyone can use these products.[46]

Approvals are granted by Ministers after taking advice from the Advisory Committee on Pesticides. Preservatives approved for use are described in the MAFF/HSE publication Pesticides 1996, and subsequent annual editions.[53] The existing regulations are to be updated to clarify the law on access to information and introduce a duty to ensure that pesticides are applied only to the target.[44]

Wood Preservatives in the Dock

In 1992, Rentokil Ltd made an out of court settlement of £90,000 with George Yates, an ex employee who developed soft tissue stomach sarcoma from exposure to Lindane and pentachlorophenol wood preservatives between 1978 and 1988.[6, 48] Experts agreed that the sarcoma was caused by dioxin impurities in the PCP or by the chemical itself,[6] and Mr Yates charged that Rentokil did not provide him with adequate protective clothing and respiratory equipment.[48]

Rentokil also paid an out of court settlement to the family of a 16 year old who died of aplastic anaemia after their home had been sprayed with these pesticides.[6]

The British Wood Preserving and Damp Proofing Association point out that these claims were settled by the companies insurers in order to avoid the costs of litigation. They report that in a more recent case, Rentokil decided to fight the case in court and their defence was successful due to the failure of the

litigant to prove a link between their symptoms and exposure to preservatives. The BWPDP association were unable to cite any press reports on this case.[49]

Occupational Exposure Limits

There are suggestions that the control limits for preservatives and other pesticides are seriously deficient.

The information used to determine OCLs is derived from animal tests which establish No Observable Effect levels (NOELs) which are extrapolated to give human OCLs by applying a safety factor, which is the subject of scientific dispute.[56]

Also, no consideration is given to possible synergistic effects of the cocktails which comprise wood preservatives, which "might give rise to significantly greater risks than those expected from simple addition of their several actions".[58] Occupational exposure limits (OCLs) do not even recognise the simple additive effects of multiple low dose exposure, let alone any possible interactions.[56]

and feeling.[22, 29] It has also been linked with anorexia[63] plus skin and other cancers.[22]

Ingestion can lead to acute symptoms of abdominal pain and vomiting, diarrhoea and muscle cramps.[64]

Handling wet timber has resulted in arsenic poisoning, and splinters fester painfully under the skin.[22, 29, 34]

Timber is most dangerous in the first two weeks after treatment,[22] but despite its high toxicity in solution, CCA forms a non-leachable, stable compound on drying, and is a unique preservative in this respect.[33, 53]

Disposal

The American publication, Environmental Building News, is calling for a ban on CCA preservatives, due to the problems associated with the disposal of CCA treated timber.[74]

Other

Arsenical chemicals impart brittleness to the wood, causing excessive wear and splitting in high stress situations.[25] CCA corrodes some metals, and so corrosion resistant fixings should be used.[30, 31]

CCA also has a limited decorative effect, imparting a green colour which turns grey on exposure to light.[31]

(c) Chromium Salts (Also see Arsenic above)

The worst hazards are from Chromates and Dichromates (the salts of hexavalent chrome).

Health

Water soluble chrome compounds are extreme irritants and highly toxic.[69] The main risk of acute poisoning is through swallowing or inhalation. Can cause allergic skin rashes and increase the risk of lung cancer. Dusts, liquids

Disposal

Some two million tonnes of timber are disposed of annually, much of which will contain preservative chemicals. While there is some scope for re-use, most of this will be disposed to landfill, or burned.

Eventual decay within a landfill will release a cocktail of preservative components into a liquid leachate, which is of concern as there is a general acceptance that all landfill liners will leak to some extent,[68] leading to potential for land and water contamination.

Waste wood can successfully be used as a fuel for heating and power generation - but there is concern over the emission of toxics combustion products of preservatives.[5] These will include heavy metals such as chrome and arsenic from CCA, and dioxins, furans and other organic toxins from the combustion of chlorinated preservatives such as PCP and lindane. Treated timber should NEVER be burned on domestic fires,[29] although it is often impossible to tell if a timber has been treated with toxic preservatives. As a precaution, we recommend no construction timber should be burned on domestic fires.

Disposal of pesticides has also led to numerous pollution incidents and a gradual build-up in the environment. Although most incidents are caused by agricultural use and use of preservatives on boats, those in the construction industry should use preservatives carefully, disposing of any excess or waste through reputable licensed waste disposal contractors.[7]

The Land Contamination Legacy

The manufacture and application of timber preservatives has left a legacy of land contamination. In the USA alone, some 250 sites previously used for timber preservative application are in need of some degree of remediation due to contamination of the soil with creosote, polycyclic aromatic hydrocarbons, dioxins and furans.[16]

Fungi Bite Back!

Cleanup of a PCP contaminated site in Southern Finland using fungus has been carried out by Cardiff based firm Biotal at a cost of £100 per tonne of soil. The fungus, Phanerochaete chrysosporium, reduced PCP levels from 700mg/kg to 10mg/kg within a year, and destroyed dioxins, PCBs and PAHs present in the soil.[55]

Using conventional excavate and incinerate technology would have cost £1000 per tonne of soil.[55]

Water Pollution

Pollution of land and water is generally from poor disposal of excess preservatives or spillage from outdated plant that does not meet existing regulations or codes of practice. Such incidents arise as a result of leakage from the treatment plant, runoff from freshly treated timber or problems arising from the storage of sludge.[53] For example, Hickson International Plc (Vac-Vac and Tanlith treatments) were recently convicted of 21 instances of illegal discharges of trade effluent and chemicals into the River Aire, West Yorkshire.[59]

Discharge of preservatives could be expected to have damaging effects on marine and terrestrial ecology, even at low levels.[56]

and vapours are irritants and corrosive to the skin, and can cause ulceration of the skin and nasal septum (the wall between the nostrils).[22]

Environment

Chrome compounds are toxic to aquatic organisms, and levels of chrome in water are controlled by a series of environmental quality standards.[36]

(d) Copper Salts (also see Arsenic above)

Health

The metal itself seems to cause few health problems unless heated to produce fumes - which can induce flu-like symptoms.[22]

Copper sulphate is toxic if swallowed and severe gastric disturbance, damage to liver and kidneys and to the nervous system have been reported in acute poisoning. It is also reported to cause eye and skin irritation.[22]

Copper naphthanate is a severe eye and skin irritant.[22]

Copper quatermium is reported to be a less toxic alternative to copper chrome arsenate.[72]

(e) Fluorides

Three compounds, ammonium bifluorate, potassium bifluorate and sodium fluorate are used in wood preservatives.[22]

Health

Fluorides may cause irritation to the skin, eyes and respiratory system, with the possibility of allergic reaction. The acute effects of overexposure include nausea, abdominal pain, diarrhoea, thirst and sweating. The main risks from chronic exposure are kidney damage and brittle bones.[22]

9.4.4 Synthetic Organic Preservatives

These are potentially the most dangerous active ingredients because they are volatile in timber and persistent in the body and in the environment.[56]

(a) Dieldrin

Cited as one of the most dangerous preservatives,[7,8] dieldrin now has no approved preservative uses and was banned in the UK in 1992 following EC legislation.[22] According to Shell Chemical, the sole producer of dieldrin, world production ceased in 1991.[51]

Health

Dieldrin is classed as highly poisonous, the lethal dose in adults around 2-3 grams. It poisons through the skin acting as a nerve poison and carcinogen.[22, 29]

Fire & Explosion

See 'solvent borne preservatives', p.106

Disposal

Incineration of timber treated with high-chlorine preservatives such as dieldrin leads to the production of extremely toxic dioxins and furans.[50]

(b) Pentachlorophenol (PCP)

Previously used in pretreatment and for remedial & DIY preservatives,[22] the use of PCP is now severely restricted under EC Directive 76/769/EEC.[33, 41,53]

Denmark, Sweden, Finland and Germany have banned the use of PCP entirely,[45,53] and in the USA it is restricted to professional outdoor use.[22, 34,]

Health

Classed as highly poisonous,[22] blamed for 1000 deaths worldwide, PCP is more poisonous through the skin than other routes. Wood, air and objects in treated buildings remain toxic for years.[22]

Acute (short exposure high dose) and chronic (long exposure, low dose) poisoning may occur by absorption through the skin, inhalation or ingestion.[11,15]

Chronic exposure has been shown to result in skin symptoms such as rashes[10,15,17] (pemphigus vulgaris, chronic urticaria and chloracne).[10, 17] Chloracne has been found to be endemic in factories where PCP is manufactured.[17]

Chronic exposure also has effects on the respiratory and nervous systems, the kidneys, digestive tract and metabolism, the latter leading to raised temperature and fever.[15] Although not classified as a human carcinogen,

---

BATS

Preservatives have had a catastrophic effect on populations of bats in Britain, which have relied on roosting in eaves of roofs for hundreds of years.[34] Bats use buildings of all types and ages as summer roosts, causing no harm whatsoever to the building.[29]

Bats die if they roost in lofts treated with dieldrin, lindane, pentachlorophenol or TBTO and heavy fines can be levied for using these chemicals where bats roost.[22]

If signs of bats are discovered then the law requires that work be halted until English Nature's (or equivalent body) Local Officer has inspected the roost and given advice.[29]

Information on bats can be obtained from your local wildlife trust, or from the Vincent Wildlife Trust, who produce a factpack aimed at the building trade, entitled 'Bats in Buildings' (Vincent Wildlife Trust, Paignton Zoo, Totnes Rd, Paignton, Devon TQ4 7EU. (01803) 521064)

some observations suggest that exposure to chlorophenols in general, and PCP solutions in particular may lead to an increased risk for certain malignant disorders such as nasal carcinoma and soft tissue sarcoma.[15]

Research suggests that chlorophenols may act as 'promoters' or cocarcinogens, and the immune system is particularly sensitive to their toxic effects. Transfer to unborn children across the placenta may result in embryotoxicity and abortion.[11]

Several studies have found PCP to be carcinogenic to animals[13,14] and a 1991 study suggested that the carcinogenic effect was exclusively due to PCP itself, with the possibility of a minimal potentiating influence by contaminants.[13]

PCP irritates the eyes, nose and throat, leading to sneezing & coughing. Weakness, weight loss, sweating, dizziness, nausea, vomiting, breathlessness, chest pain, dermatitis,[22] headaches, sleeplessness, lack of concentration and psychosomatic problems are reported to be caused by PCP and Lindane.[21]

No specific antidote exists for the treatment of acute PCP poisoning.[15]

A 1993 study into PCP poisoning carried out at the University of Antwerp concludes that the use of PCP based products as indoor wood preservatives poses an unacceptable risk to human health.[15]

Chronic poisoning occurs mainly to sawmill workers exposed to preservative contaminated sawdust, and those living in homes containing a large amount of PCP treated timber.[15]

A 1986 study showed that all occupationally exposed groups sampled, including timber yard workers, preservative formulation workers and those involved in applying preservatives, showed evidence of substantial PCP absorption. The highest absorptions were found in remedial timber treatment operatives.[12,14]

Net daily intake of PCP by people not occupationally exposed, in eight countries varied from between 5 micrograms (Nigeria) to 37 micrograms (Netherlands). In individuals occupationally exposed, daily intake ranged from 35 micrograms to 24,000 micrograms, depending on the type of work.[14]

PCP also contains impurities such as dioxins, which are also highly toxic.[22]

Environment

Rhone-Poulenc are the last European manufacturers of PCP, although PCP is no longer produced in EC countries.[45] Waste from their manufacturing plant in the forest region of Cubanto, Brazil, the London Hazards Centre reports, are dumped in the forest, leading to contamination of ground and drinking water and damage to the forest. This in turn has lead to serious health problems in the local community.[22]

PCP is a marine pollutant, found at up to 20ppm in marine sediments, where its toxicity poses a risk to aquatic organisms.[45]

Residues of chlorophenols have been found worldwide in water, air and soil samples, in food products and in human and animal tissues and body fluids.[11] Wood preservative chlorinated phenols contribute to this contamination mainly through industrial effluents and poorly disposed of waste preservatives.

Chlorophenols are found ubiquitously in the environment. Mono-, di-, tri- and tetrachlorophenols as well as pentachlorophenol occur in the urine of the general population often in 'surprisingly high concentrations'.[18]

Fire & Explosion

See Organic Solvent Borne Preservatives section, p.106

Disposal

Incineration of timber treated with high-chlorine preservatives such as PCP leads to the production of extremely toxic dioxins and furans.[50]

Other

PCP imparts brittleness to wood, causing excessive wear and splitting.[25]

Greenpeace are calling for an end to all chlorine chemistry due to the danger it presents to human health and the environment.

(c) Lindane (Gamma-HCH)

Restricted to professional use by licensed contractors.[33]

Health

Classed as highly poisonous,[22] lindane has been implicated in illness and death of several people after their homes had been treated for woodworm.[34] Lindane has been banned or severely restricted in Japan, USA and many other countries.[22]

Upper airway irritation, dry mucosa, headaches, sleeplessness, lack of concentration, psychosomatic problems, cyanosis, aplastic aneamia, muscular spasms

---

'PEGS'

People affected by exposure to wood preservatives, or other forms of pesticide, can contact PEGS - the Pesticide Exposure Group of Sufferers. PEGS can offer practical advice and counselling.

Many firms have recently changed their chemicals for safer formulations, and so the majority of PEGS's callers tend to be DIY'ers who are using out-dated products which they may have had stored for several years, without due regard to the instructions.[6]

Recent studies suggest that up to one in eight people may be sensitive to the low levels of preservatives such as lindane, TBTO and PCP found in treated timber in the home.[54]

PEGS can be contacted on 01223 64707 or 01766 512548

and convulsions are reported as caused by the wood preservatives PCP and Lindane.[21,22,29]

Lindane has an oestrogenic effect (ie, mimics the hormone oestrogen), known to encourage the growth of breast cancers[42] and research published in 1993 implicated lindane as a cause of childhood brain cancer.[57]

The organs affected are the eyes, central nervous system, blood, liver, kidneys and skin.[22]

Environment

The use of lindane in roof spaces has killed many colonies of bats and has been one of the main reasons for their decline in Britain over the last few decades. Lindane can remain at lethal concentrations to bats for more than 20 years after treatment.[34]

Lindane is found in the North Sea, and is a ''serious problem'' as a marine pollutant.[40]

Fire & Explosion

See Organic Solvent Borne Preservatives, p.106

Developing Countries

It is reported that some suppliers and government institutions and even some recent publications operating in developing countries, still recommend the use of preservatives and other pesticides such as PCB (Polychlorinated biphenyls) and PCT (Polychlorinated terphenyls) which have been banned in industrialised countries.[7,9]

No chemical preservative should be used without full knowledge of its composition, and is is recommended that preservatives containing DDT, PCP, Lindane and arsenic SHOULD BE AVOIDED.[9]

Many preservative and other pesticide manufacturers, unable to meet stringent US and European safety standards, have relocated to developing countries such as Brazil and Mexico where poor environmental controls at treatment plants have led to widespread pollution.

Dioxins

PCP and lindane are known to contain dioxins,[21] and a 1995 study suggests the possibility of a suppressive effect on the immune system of building users by dioxins released from preservative treated wood.[20]

The carcinogenic potency of the most thoroughly studied dioxin, 2,3,7,8-TCDD, is more than 140,000 times greater than that of lindane and 7,800 times that of dieldrin.[52]

There is no known safe level for dioxin exposure.[56]

Paints & Stains

Many types of wood stain contain fungicides, mainly to reduce disfigurement by staining fungi. If children or animals are to come into contact with woodwork finished with a product containing a fungicide, BRE recommend that you seek a specific assurance on safety from the manufacturer or consider using an alternative product instead.[31]

Disposal

Incineration of timber treated with high-chorine preservatives such as lindane, leads to the production of extremely toxic dioxins and furans.[50]

Other

Lindane is reported by the BRE to be not as effective as pyrethroid compounds, which are also less toxic.[22]

Greenpeace are calling for an end to all chlorine chemistry due to the danger it presents to human health and the environment.

(d) Dichlofluanid

A fungicide.

Health

A fungicide of moderate oral toxicity, dichlofluanid can cause skin and eye irritation. There is some evidence of mutagenicity in laboratory microbial tests.[46]

Environment

Dichlofluanid is harmful to fish.[46]

Fire & Explosion

See Organic Solvent Borne Preservatives, p.106

Disposal

Incineration of timber treated with high-chorine preservatives such as Dichlofluanid, leads to the production of extremely toxic dioxins and furans.[50]

Other

Greenpeace are calling for an end to all chlorine chemistry due to the danger it presents to human health and the environment.

(e) Tributyl Tin Oxide (TBTO)

Restricted to professional use by licensed contractors.[33]

Health

TBTO affects the central nervous system, eyes, liver, urinary tract, skin and blood.[22]

The symptoms of poisoning are headaches, vertigo, eye irritation, psychologic neurological disturbance, sore throat, cough, abdominal pain, vomiting, urine retention, paresis (slight paralysis/weakness), skin burn and pruritus (itching).[22]

TBTO has been shown to damage the foetuses of animals, but the effects on human reproduction are unknown.[22] The cancer risk and damage to the immune system are still under investigation.[22]

Environment

Banned as a boat antifouling paint (for boats under 25 metres in length) in 1987[37] due to its effects on marine life, including commercial mussel beds.[22,34] Effects on shellfish include deformities and sterility, which can be caused at concentrations as low as 5ng/litre (0.000 000 005g/litre)[39] Shellfish populations in parts of the North Sea are still suffering from exposure to TBTO,[37] although populations of mussels are shown to have increased since the ban.[34] A complete ban on TBTO is being considered.[39]

(f) Permethrin

An insecticide developed in 1973, modelled on a natural insect poison derived from the daisy species *Chrysanthemum cinerariaefolium.*

Health

Permethrin is moderately toxic in its undiluted form,[46] and the human lethal dose is high - around 35 grams.[22] Natural pyrethroids have a low oral toxicity to mammals but are expensive.[63,64] Synthetic pyrethroids have a wide ranging toxicity, from the relatively non-toxic bioresmethrin to the highly toxic decamethrin. Dermal toxicity is considered so low as to not be considered a hazard.[64]

Although generally considered safer and less persistent than many of the alternatives such as lindane[22,46,53] (permethrin is estimated to be one fortieth as toxic as lindane),[56] permethrin has been the centre of an acrimonious dispute in the US, with some researchers at the Environmental Protection Agency suggesting that it may be a carcinogen. The carcinogenic risk from DIY use however, is likely to be small[34]

Associated with allergy, irritation,[22] conjunctivitis[46] and nervous system damage,[22] particularly the nerve control of breathing and muscles in acute cases.[46]

Environment

Generally considered to be far safer to bats and other wildlife, and is recommended by the Nature Conservancy Council in their advice leaflets about protecting bats.[34] However, permethrin is highly toxic to fish if it drains or washes into watercourses.[46]

(g) Copper & Zinc Naphthenates

Claimed to have a lower mammalian toxicity than traditional materials.[33,53]

The BWPDPA accept that zinc formulations are as effective as the more dangerous preservatives which are covered by integrated pollution control (IPC) legislation, if 15-20% more expensive.[36]

(h) Acypetacs-zinc

A fungicide with insecticidal properties. Claimed to have a lower mammalian toxicity than traditional materials,[33] this is still a relatively new compound (first reported in 1983[46]) and little is known about its long term health and environmental effects.[46]

Useful Contacts

The London Hazards Centre:	0171 837 5605
The Pesticides Trust:	0171 274 8895
The British Wood Preserving and Damp Proofing Association:	0181 519 3444
HSE Information Centre:	0114 289 2345

"Even the most rigorous chemical treatment provides little more than temporary protection of structural timber, and the long term guarantees commonly offered are in our view entirely speculative"

Hutton & Rostron Environmental Investigations Ltd.

9.5 Alternatives

9.5.1 Alternative 'chemicals'

Preservatives based on boron (see Product Analysis, p.107), soda, potash, beeswax and linseed oil are recommended as safe,[9] and although not as rot resistant as their poisonous counterparts, can be equally effective if used in conjunction with good building design.[9]

Coating timber with raw linseed oil is recommended as a safe, traditional preservation method.[46] The drawbacks are that this gives a sticky finish which tends to hold dirt and supports the growth of staining fungi,[31] although this only causes superficial rather than structural damage.

9.4.2 Avoiding Preservatives Altogether

Eradication of biological agents from buildings is impossible, and environmental control and preventative maintenance should make 'draconian' chemical treatments unneccessary,[66] providing a long term solution to the health of buildings and their occupants.[62]
There is NO REASON AT ALL to use preservatives on non-structural and decorative timbers.

(a) Naturally Durable Wood

Some timbers such as cedar have been valued for their pesticidal effect for thousands of years. Similarly, resins and tannins have been extracted from timber to produce toxic wood creosotes and wood pitches - although these can be as toxic to humans as their coal derived counterparts, and their effectiveness is questionable.[61]
Although wood of suitable natural durability can be specified as an alternative to preservative treatment, there can be cost penalties, engineering problems and difficulties in supply.[4,9] Also, many of the more durable species are tropical hardwoods, presenting a conflict with the environmental concerns for their conservation and more rational use.[4]
However, specification of durable timbers from certified sustainable sources reduces impacts, and supports sustainable forestry (see chapter 7).

(b) Seasoning

Properly seasoned wood has a natural resistance to decay organisms. Seasoning involves stripping the log and soaking it to remove sugars from the sap, which act as a food source to organisms responsible for decay. Soaking can be achieved by floating the logs in a river or pool, or by spraying.[8]
The wood must then be dried to its equilibrium water

content (8 - 20% by weight, depending on the timber species and climatic conditions)[9]. Natural drying is preferable in terms of energy and quality of the timber, but kilning is often used to reduce drying times, although this could be regarded as a waste of energy, and can result in warping and shrinkage.[8]

(c) Environmental Control

Traditionally, timber decay was controlled through proper seasoning and careful detailing. The recent reliance on cheap preservative treatments to control rot and infestation is blamed for a general slackening in good seasoning practices and design that protect timber naturally.[8] An increase in the sealing of buildings for energy conservation has contributed to this by increasing the risk of condensation.[8]

Insect and fungal attack will occur mainly in just two situations; in poorly dried wood, and wood which becomes wet, and is unable to dry out due to poor ventilation, enclosure within the building structure or by impervious protective finishes.[56]

The "environmental" control of timber decay involves control of decay organisms through regulating temperature and available moisture.[29] Correction of environmental conditions will generally prevent further developments of insect pests and rot as neither dry nor wet rots can survive in dry, well ventilated conditions.[46] In simplistic terms, environmental control involves two principles;

1. Correct building faults leading to high moisture content in timber
2. Increasing ventilation around timber at risk.[29]

Blocking the moisture source with impermeable materials is generally ineffective, and may lead to high moisture content and decay in adjacent materials.[29]
The most effective method is to balance every moisture source with a moisture sink - eg, venting moisture from occupation out through windows and chimneys[29] and extractor fans in damp areas such as kitchens and bathrooms.[62]

Common Examples - Sources & Sinks

Damp Proof Course Bridging
Often occurs as a result of raised ground levels, leading to ingress of moisture and decay of timbers in adjacent floor spaces.
Reducing the ground level will remove the moisture source and provide a moisture sink by allowing evaporation from the exposed wall. Additional moisture sinks such as air-bricks or sub-floor ventilation are also common measures.[29]

Roof/gutter/coping failure
Can result in significant water ingress into the masonry beneath, which will then act as a moisture reservoir. Any timber in contact with this will tend to 'wick' moisture from the masonry. The moisture source can be eliminated by roof repair - and its effectiveness as a source eliminated by isolating in-contact timber from the masonry using dpc membranes or joist hangers to produce an air gap. It must also be ensured that timbers are adequately ventilated.[29]

Closed Cavities & Water Impermeable Layers
Bricked up lintels, and sealed up emulsion painted sash windows are examples of situations where moisture can be trapped into the wood, leading to a risk of rot.[29]
It has been demonstrated that decay can be arrested in some cases simply by reducing the moisture source, drying out the wood and increasing ventilation - without the use of toxic treatments[8] If in doubt, specialist advice should be sought.
Most of the remedial work required to remove moisture sources/provide sinks are traditional repairs, well within the capacity of the general building contractor. New products and materials such as time controlled fans, hollow ventilation plastic skirting boards, plastic masonry grains, roof space ventilation systems, moisture permeable paints, dry lining, joist hangers and tanking, can also be useful.[29]

(e) Replacement

If rot is so advanced that the timber needs replacing, it is important to choose well seasoned timber and install it in such a way as to ensure that it will remain reasonably dry. The best way of achieving this is to provide ventilation around the timber if it is situated near a potential moisture source.[8] Timber affected by rot should be cut away to one metre beyond the last sign of decay before being replaced.[46]

(f) Detailing

Windows, doors and door frames are the items most commonly pre-treated as standard by the large manufacturers - but the requirement for this can be removed by good detailing. For example, frames can be made up with all rebates inclined, and alloy free-draining bottom beads to prevent water entrapment. Treatment of bottom rails and sills may still be required in exposed situations, and a thorough paint or stain system is essential.[56]
Common faults in windows which lead to decay are shrinking of glazing putty away from the glass, dry bedded beading, and gaping joints.[67]

Further information of window frame detailing can be found in Chapter 10.

(g) Physical Barrier Control Methods

Termites & Other Insects

Insects can be a particular problem in the tropics and sub-tropics, where a number of non-chemical measures are recommended.

A continuous reinforced concrete floor slab can effectively keep out subterranean termites. If joints are necessary, these should be rough and sloping or tongue and groove joints.[9]

A sand barrier around a buildings foundations can prevent termites from reaching a building[65] - for design details, see the March/April 1994 issue of Environmental Building News.

Buildings raised 80-100cm off the ground on poles or columns permit visual inspections underneath the floor and also facilitate ventilation. Painting exposed foundations and columns a light colour will make termite galleries more visible from a distance.[9]

New methods of termite control are currently under development in Germany, involving cross breeding and elimination of the reproductive capacity of termites, production of sexual hormones to disorient the termites, or alarming pheromones and repellents to start an escape reaction.[9]

If toxic treatments must be used (eg, for Professional Indemnity Insurance), localised, well focused application of chemicals is recommended to prevent termites entering the building, rather than wholesale application throughout the structure.

9.5.3 Heat Treatment

(a) Pre-Treatment

Also known as 'retification', this process involves heating the timber in an oxygen depleted atmosphere to 160°C. This causes some chemical modification of the timber which appears to inhibit decay quite successfully.[5] It is believed that a commercial plant is operating in Germany.[5]

(b) In-Situ Treatment

This is achieved using hot air or microwaves to kill existing fungal infestations. Whilst safe and effective, it does not prevent reinfestation,[5] although this can be achieved by adjustment of environmental conditions as described above.

9.5.4 Monitoring and Regular Inspection

This involves the installation of monitoring systems that log the ingress of moisture into the buildings fabric. This allows precise location of faults which can then be rectified to prevent further wetting of the timbers. Such systems are not a complete substitute for preservative treatments because timbers remain at risk from decay after faults have been rectified, during the drying out period which may take several years.[5]

Such systems could be installed in new buildings along with burglar and fire alarms.[5]

Similar systems are already widely used in the USA to detect insect infestation. This allows remedial work to be carried out in good time, avoiding a need for wholesale use of precautionary preservatives and insecticides.

Such inspections are often free, presumably costed into remedial work as and when required.

As an aid to inspection, insect pheromone traps have been developed, which allow easy identification of the species present.[5]

It is also important to regularly inspect a building to ensure the continued integrity of damp proofing, water exclusion and building services so that the internal structure remains dry.

It is recommended that structural timbers are exposed within a building so that woodworm infestation shows up in time to deal with it.[72]

9.5.5 Alternative Treatments for Fungal Infection

(a) Closely Targeted Preservatives

A possible future development would be preservatives targeted at specific organisms or groups of organisms, operating by interfering with essential biochemical processes.

There are four current fields of study;

i. Wood - Fungal interaction

Fungi attack wood using 'hyphae' which penetrate the cell wall. The principal component of the hyphal membrane are compounds known as 'sterols'. A compound that could inhibit the formation of these, or otherwise disrupt the structure of function of the hyphae would stop the fungi's ability to rot wood.[5]

ii. Control of Digestion

Fungi digest wood using enzymes. Once excreted, enzymes are fairly stable, but compounds which interfere with their secretion would be effective in preventing rot. Surfactants and detergents have been suggested as a promising means of achieving this.[5]

A similar approach for the control of insect attack is focusing on the control of micro-organisms in the gut, which produce enzymes useful to the insect.[5]

iii. Control of Growth

Trials have been made on a fungus 'Junk food' amino-isobutyric acid (AIB), which can be taken up by the fungus but not utilised for growth, causing infestations to die from malnutrition![5]

Hutton & Rostron have been involved with the development of a technique using alpha-AIB as an anti-fungal treatment in paints and forestry.[55]

Further information can be obtained from:

Hutton & Rostron Environmental Investigations Ltd
Netley House
Gomshall, Surrey, GU5 9QA
Tel. 01486 413221

iv. Chelating agents

Magnesium and iron appear to play an important role in brown rot decay, and attempts have been made to limit their availability to fungi by using chelating agents.[5]

(b) Bio-control

Certain bacteria and fungi do not decay wood, but do inhibit the growth of decay fungi - either by using them as a food source or by producing antibiotics that kill or inhibit wood decaying fungi.[5]

There are currently two fungi, Peniophora gigantea (spore tablets) and Trichoderma (pellets or wettable powder), mainly in agricultural use to protect wood and trees against harmful fungi. There is potential for their use for protecting telegraph poles and marine piles,[63] and in one experiment, material removed from Trichoderma-treated poles resisted attack by active fungi for seven years.[5]

A number of parasites of wood borers exist, particularly nematode worms and parasitic fungi, but there would be problems of delivery for timber treatment.[5]

9.5.6 Recycled Plastic Lumber

One way of avoiding the use of preservatives in high risk applications such as marine pilings is to use recycled plastic lumber in place of timber. Plastic lumber is said to be totally resistant to biological decay, and may last up to 600 years![71]

Plastic lumber is covered in detail in issue 13 of the Green Building Digest Magazine.

Water Pollution from Preservatives Used in the Home

It is customary to attribute the occurrence of organochlorine pesticide residues in UK surface waters primarily to industrial and agricultural sources. However, research at the Queens University of Belfast showed as early as 1977 that organochlorines in sewage effluents appear to have a domestic origin - "probably arising from the use of wood preservatives in the home".[73]

The researchers concluded that the levels of benzehexachloride (BHC) and other pesticide residues in UK waters may well pose a threat to aquatic ecosystems.[73]

Glossary

Acute: Immediately detectable, or short term effects / high dose, short period of exposure.
Chronic: Effects in, or only detectable in, the long term / low dose, long period of exposure.
Carcinogenic: Able to cause cancer
Mutagenic: Can cause genetic damage
Teratogenic: Can cause abnormal foetal development
Synergistic Effects: Where the combined effect of two or more substances is greater than the sum of the separate effects.

9.6 Environment Conscious Suppliers

(a) Pretreatment

DRY PINS Wood Preservative Capsule

Window Care Systems Ltd, Unit E, Sawtry Business Park, Glatton Road, Sawtry, Huntingdon, Cambridgeshire PE17 5SR (tel. 01487 830311 fax. 01487 832876)

Certification: HSE No: 5144

Guide Price: £1.10 per pin

A dry system for local wood preservation of new and existing wooden constructions.

Antiflame 4050 WD Flame retardant and timber preservative

Safeguard Chemicals, Unit 6, Redkiln Close, Redkiln Way, Horsham, Sussex, RH13 5QL (tel. 01403 210204 fax: 01403 217529)

Contact: David Payne

Dual function micro emulsion flame retardant/timber preservative which claims to be "odourless, non-toxic and works as an extinguishing media." Active ingredients are Boron and permethrin.

Timbor Rods - Preservative filled glass rods

Biokil Chemicals Ltd, 14 Spring Road, Smethwick. B66 1PE (tel. 01747 823121 fax. 01747 822636)

Contact: Mike Dunn

Application for protecting heartwood and sapwood from fungal decay in both new and existing timbers. Simple to install, timbor rods are compatable with most other materials. The rod remains intact until activated by any moisture entering the timber.

(b) Remedial Treatment

BORAX No. 2031 Impregnating Preserevative

The Nature Maid Company - BIOFA UK, Unit D& Mans Craft Centre, Jackfield, Ironbridge, Shropshire, TF8 7LS (tel. 01952 883288 fax. 01952 883200)

Gary Taylor

A sodium borate rich, water based, nearly neutral solution of boric salts, it has been approved by the institute of building technology in Berlin as effective against fungi and insects. It can be brushed, sprayed or dip applied.

BORAX No. 111 Wood Preservative

Auro Paints Ltd, Unit 1, Goldstone Farm, Ashdon, Nr. Saffron Walden, Essex CB10 2LZ (tel. 01799 584888)

Contact: Richard Hadfield

Clear and odourless borax wood preservative powder

Deepwood 20 & Deepwood 50

Safeguard Chemicals, (Address Above)

Contact: David Payne

A clear odourless viscous liquid designed for liberal application to surfaces by means of brushing. For use against both fungal decays and woodboring insects. Designed to be especially effective against dry rot and the wet rots in remedial situations. They are based on water soluble borate, and suitable for high risk areas such as embedded joist ends, lintels etc

Boracel 20 Inorganic Boron Remedial Treatment

Remtox Chemicals Ltd, 14 Spring Road, Smethwick, Warley, West Midlands, B66 1PE (tel. 0121 525 5711 fax. 0121 525 1740)

Contact: Walter Howarth (Marketing)

Remtox use compounds developed from chemicals with low mamalian toxicity, and recommend the use of non-solvent products wherever possible.

Contractors are required to send personnel for training, and are inspected on a regular basis by fully trained technical representatives.

Renlon Group Ltd (Various preservatives), 148 South Park Road, Wimbledon, Greater London, SW19 8TA (tel. 0181 542 9875 fax. 0181 542 7007)

Awards: Marley Environmental Award, 1993

The companies long term aim is to use 'zero toxic chemicals' while protecting properties against insect infestation and fungal decay.

'Timbor' Timber preservative

Borax Consolidated Ltd, 170 Priestly Road, Guildford, Essex GU2 5RQ (tel. 01483 734000 fax. 01483 457676)

Contact: Jeff Lloyd

Timbor contains borates, with no other added formulations. Formulated to give the highest concentration of boron at the maximum possible solubility. Applied as aqueous solution, and therefore doesn't require solvents. Can be applied by vacuum pressure process to dry or green timber or by dip diffusion to green timber.

Listings supplied by the Green Building Press, extracted from 'GreenPro', the interactive building products and services for greener specification database. At present, Greenpro lists over 600 environmental choice building products and services throughout the UK and is growing in size daily. The database is produced in collaboration with the Association for Environmentally Conscious Building (AECB).

For more information on access to this database, contact Keith Hall on 01559 370908 or e-mail buildgreen@aol.com

9.7 References

1. Pre-Treated Timber: Environmental and Energy Benefits (G. Ewbank) in: The European Directory of Sustainable and Energy Efficient Building 1995 (O. Lewis & J. Goulding, Eds). James & James Science Publishers, London

2. The Construction Materials Reference Book. (D.K. Dorran, Ed.) Butterworth-Heinmann, Oxford, 1992

3. Mitchell's Materials, 5th Edition (A. Everett & C.M.H. Barritt) Longmann Scientific & Technical, UK. 1994

4. An Environmental Assessment for New Homes. BREEAM/ New Homes version 3/91 (J.J. Prior, G.J. Raw & J.L. Charlesworth). Building Research Establishment (BRE). 1991

5. Wood Preservation for a Sustainable Future. (Gervais Sawyer). Paper presented to the Sustainable Futures conference, Hull, 2nd February 1996

6. Building for a Future Vol.2 No.4 Winter 1992/93

7. Buildings & Health - The Rosehaugh Guide. (S Curwell, C March & R Venables). RIBA Publications, London. 1990

8. Eco-Renovation - The Ecological Home Improvement Guide. (E. Harland) Green Books Ltd, Devon. 1993

9. Appropriate Building Materials, 3rd Edition (R. Stulz & K Mukerji). SKAT Publications & IT Publications, Switzerland, 1993

10. Skin lesions as a sign of subacute pentachlorophenol intoxication. (J Lambert, P Schepens, J Janssens & P Dockx). Acta. Derm. Venereol. (Stockholm) 66 (2) p.170-172 1986

11. A Review of Chlorinated Phenols. (J H Exon). Vet. Human Toxicology (USA) 26 (6) p508-520. December 1984

12. Absorption study of pentachlorophenol in persons working with wood preservatives.(RD Jones, DP Winter & AJ Cooper) Human Toxicology (England) 5 (3) p189-194. May 1986

13. Toxicology and Carcinogenesis studies of two grades of pentachlorophenol in B6C3F1 mice. (EE McConnell, JE Huff, M Hejtmancik, AC Peters & R Persing). Fundamental. Applied Toxicology (USA) 17 (3) p519-532. Oct 1991.

14. Assessment of pentachlorophenol exposure in humans using the clearance concept. (BG Reigner, FY Bois & TN Tozer). Human Experimental Toxicology (England) 11 (1) p17-26. Jan 1992.

15. Human Pentachlorophenol Poisoning. (PG Jorens & PJ Schepens). Human Experimental Toxicology (England) 12 (6) p479-495. Nov 1993

16. Comparing the results of a Monte Carlo analysis with EPAs reasonable maximum exposed individual (RMEI): A case study of a former wood treatment site. (TL Copeland, DJ Paustenbach, MA Harris & J Otani). Regul. Toxicol. Pharmacology (USA). 18 (2) p275-312. Oct 1993

17. Chloracne. Some recent issues. (PJ Coenraads, A Brouwer, K Olie, N Tang). Dermatol. Clin. (USA) 12 (3) p569-576. July 1994

18. Chlorophenole im Harn als umweltmedizinische Untersuchungsparmeter. (R Wrbitzky, J Angerer, & G Lehnert) Gesundheitswesen (Germany) 56 (11) p629-635. Nov. 1994.

19. Potentiation of 2,6-dinitrotoluene toxicity in Fischer 344 rats by pretreatment with coal tar creosote. (RW Chadwick et al) J. Toxicol. Environmental Health (USA), 44(3) p319-336. March 1995.

20. Effects of inhalative exposure to dioxins in wood preservatives on cell-mediated immunity in day-care centre teachers, (N Wolf & W Karmaus). Environ. Res. (USA) 68(2) p96-105. Feb 1995

21. Bei "Chronisch-rezidivierenden Infekten des Atemtraktes" differentialdiagnostich Holzschutzmittelintoxikation erwogen? (English abstract) (P Ohnsorge) Laryngorhinootologie (Germany) 70(10) p556-558. Oct. 1991

22. Toxic Treatments - Wood Preservative hazards at Work and in the Home. The London Hazards Centre. November 1988

23. Building trades journal 1985, ref. in 22.

24. United Nations Environment Programme, IRPTC Bulletin, Vol. 9 No. 1. June 1988

25. NIOSH Technical Report, 1983

26. BWPA News, October 1988

27. MNWR Volume 13 p.170-171 (in ref. 22)

28. The Relative Amount of Pentachlorophenol Volatilisation from Treated Wood. (LL Ingram, GD McGinnis & LR Gjorvic) American Wood Preservers Association, 1981.

29. Greener Building Products and Services Directory. (K. Hall & p. Warm) AECB, 1993.

30. Wood Preservatives: Application Methods. BRE Digest 378, January 1993

31. Natural Finishes for Exterior Wood. BRE Digest 387, October 1993.

32. Long-term field trials on preserved timber out of ground contact (revised to 1990). BRE Report BR249. 1993

33. Going Green - The Green Construction Handbook. (Ove Arup & Partners) JT Design Build Ltd, 1993

34. The Pesticides Handbook, 9P. Hurley, A. Hay, & N. Dudley) Journeyman Press, London. 1991

35. ENDS Report 240 p.14 January 1995

36. ENDS Report 238 p.32 November 1994

37. ENDS Report 232 p.30 May 1994

38. ENDS Report 229 p.11, p.30 February 1994

39. ENDS Report 227 p.7 December 1993

40. ENDS Report 227 p.39 December 1993

41. ENDS Report 209 p.38 June 1992

42. ENDS Report 241 p.8 February 1995

43. ENDS Report 250 p.36 November 1995

44. ENDS Report 256 p.38 May 1996

45. ENDS Report 254 p.49 March 1996

46. Timber Treatments. The Pesticides Trust, London. (Undated)

47. The Pesticides Trust Review 1995

48. Global Pesticide Campaigner 2 (4) November 1992

49. Personal Communication, Mike Bromley, BWPDPA.

50. Global Pesticide Campaigner 2 (1) February 1992

51. Global Pesticide Campaigner 1 (2) January 1991

52. Dirty Dozen Campaigner. Pesticide Action Network International, June 1990

53. Environmental Impact of Building and Construction Materials, Volume E: Timber & Timber Products. (J. Newton & R. Venables) CIRIA, June 1995

54. Preservatives can make your life a misery. (G. Prior) Construction News, 3rd March 1994

55. Building for a Future, March 1991

56. Building for a Future 4 (4) Winter 1994/95

57. Archives of Environmental Contamination & Toxicology, Vol 24, 1993

58. Dr Badsha, Environmental Consultants and Analysts (ECAL) Southpost, Letter, 19th Sept. 1994 (Cited in 56)

59. The Guardian, 2nd November 1994

60. Green Design (A. Fox & R. Murrell) Longman Group UK Ltd, 1989

61. The Environmental Control of Timber Decay (TC Hutton, H Lloyd & J Singh) Structural Survey, 10 (1) Summer 1991

62. Greener Buildings - The Environmental Impact of Property. (S Johnson). Macmillan Press, 1993

63. The Pesticides Handbook (P. Hurst, H. Hay & N. Dudley) Journeyman press 1991.

64. Pesticide Poisoning - Notes for the Guidance of Medical

Practitioners. Department of Health & Social Security 1983.

65. National Parks Service (USA) Sustainable Design & Construction Database, Section 06300 - Wood Treatment. October 1995.

66. In-Situ Timber Treatment Using Timber Preservatives - Health, Safety and Environmental Precautions. HSE Guidance Note GS 46, June 1989.

67. Wood Windows - Arresting Decay. BRE Defect Action Sheet DAS13 January 1983 (updated February 1985).

68. ENDS Report 250 p.32, November 1995

69. Dictionary of Environmental Science & Technology (Revised Edition) (A Porteous) John Wiley & Sons, 1992

70. Personal Communication, Enwys Chapman, PEGS. June 1996.

71. Environmental Building News 2 (4) July/August 1993.

72. Personal Communication, Professor Christopher Day.

73. BHC Residues of Domestic Origin: A Significant Factor in Pollution of Freshwater in Northern Ireland. (D. Harper R. Smith & D. Gotto). Environmental Pollution (12) 1977

74. Environmental Building News 6 (4) March 1997

Window Frames 10

10.1 Scope of this Chapter

This chapter looks at the environmental impact of the main options available in the market for window frames. These include timber, steel, aluminium and uPVC. The impact of painting and maintenance has also been taken into account.

10.2 Introduction

Durability is perhaps the key issue for many in the choice of window frames. Wood windows have gained an undeserved reputation for being liable to rot, and the alternatives such as aluminium or uPVC are often seen as the only durable options. Wood's poor reputation can be put down to a number of factors. Cheaply made windows, badly installed, used as formers for openings while building brickwork, and only painted with primer on hidden surfaces, are off to a bad start. If maintenance is skimped on, then perhaps it is no wonder they rot.

But the processes by which such decay may become initiated, and methods for avoiding this through good design etc. are well documented and understood. Section 10.5 ('Design') on page 129 explains some of the basics of this. Well designed, well made and well maintained timber windows can and do last the lifetime of the building they are installed in.

uPVC windows are the most common replacement windows, but for new build housing, softwood remains more usual. This is mainly due to the low initial price of low-specification wood windows, and also to tradition on the building site.

Timber joinery manufacturers are beginning to respond to the challenge of uPVC windows by producing factory finished windows with the best features common on plastic windows - espagnolette locks, draught stripping, friction hinges, security beading and accommodation for wider double-glazing units.[1]

This is the Rainforest

CO_2 LEVELS ...that got chopped down

CO_2 LEVELS ...to make the charcoal

CO_2 LEVELS ...to smelt the iron

CO_2 LEVELS ...that was put in a boat

CO_2 LEVELS ...and sent abroad

CO_2 LEVELS ...to be turned into steel

CO_2 LEVELS ...that made the windows

SAVE THE AMAZON

...that went in the house that Jack built.

10.3 Best Buys

- Durable temperate hardwoods such as oak, which can, with suitable protection in design, be used without painting or preservatives, have got to be the greenest option for window frames, especially if sourced from well-managed forestry operations.

- Although certified well-managed timber is the best option for window frames, the paint type used is equally crucial: well-managed timber painted with conventional synthetic solvent-borne paints may well have a greater environmental impact over its life-time than uncertified timber maintained with a plant-based paint.

- The combination of uncertified tropical hardwood frame and a conventional synthetic paint or stain treatment may actually have a greater environmental impact than uPVC.

PVC Window Paint

Dulux have recently developed a uPVC window paint, recommended for use every 6 years.

This is surprising considering that one of the major selling points of uPVC window frames is their durability and low maintenance.

One can only conclude that unless Dulux are marketing a product for which there is no practical use, then uPVC windows require a similar maintenance programme to their wooden counterparts. Painting will also increase the environmental impact of uPVC windows over their lifespan.

Overall Environmental Impact 'Score' for Window Frames

The Product Table for Window Frames has rather more detail than a usual Green Building Digest table, and because this might make it slightly harder to read 'at a glance', we have included here a graphical representation of the overall 'scores'. This is very much a simplification of the issues, and intended really only as a bit of fun.

(It is not mathematically sound to simply add up a series of ranking scores as we have done here. There are no 'units' of environmental impact. We have not applied any weighting to the different issues, but have counted the frame material as being twice as important as the paint type.)

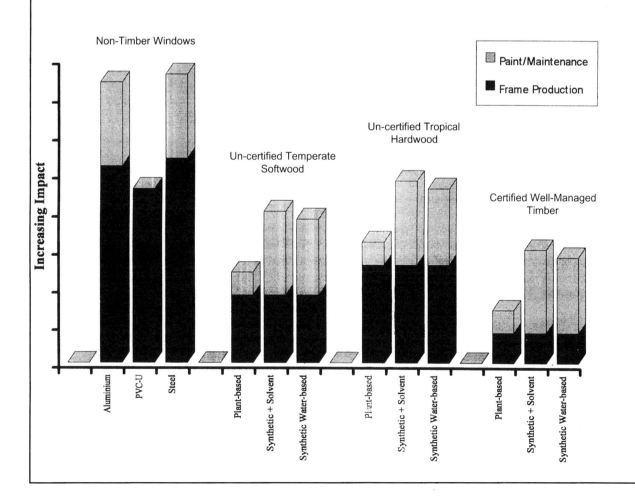

	£	Production																Use			ALERT!
	Unit Price Multiplier	Energy Use		Resource Depletion (bio)		Resource Depletion (non-bio)		Global Warming		Toxics		Acid Rain		Photochemical Oxidants		Other		Thermal Performance	Health Hazards	Recycling/Reuse/Disposal	
Rating Applies to:	Frame & Paint	Frame	Paint	Frame	Paint	Frame	Paint	Frame	Paint	Frame	Paint	Frame	Paint	Frame	Paint	Frame	Paint	Frame	Frame & Paint	Frame & Paint	
Non-Timber Windows																					
Aluminium + Paint	2.4	●	●			·	●	●		●	●	·	·	·	●	●	●	●	·	●	
u-PVC	1.0	●				●		●		●		●		●		·		●	·	●	Greenpeace Chlorine Campaign
Steel + Paint	2.3	·	●	·		·	●	·		●	●	●	·	●	●	·	●	●	·	●	
Un-certified Temperate Softwood																					
+Plant Based Paint	2.4	·	·	●					·		·			●	·			·	●	·	
+Synthetic Solvent-borne Paint	2.0	·	●	·				●	·		·			●	·	●		·	●	·	
+Synthetic Water-borne Paint	2.1	·	●	·				●	·		·		●			●		·	●	·	
Un-certified Tropical Hardwood																					
+Plant Based Paint	3.1	●	·	●				●		●				●	·			·	●	·	Friends of the Earth boycott of Brazilian mahogony etc
+Synthetic Solvent-borne Paint	2.7	●	●	●				●	●	●		·		●	·	●		·	●	·	
+Synthetic Water-borne Paint	2.8	●	●	●				●	●	●			●	●	·	●		·	●	·	
Certified Well-Managed Timber																					
+Plant Based Paint	?	·	·							●				●	·			·	●	·	
+Synthetic Solvent-borne Paint	?	·	●					●		●		·		●	·	●		·	●	·	
+Synthetic Water-borne Paint	?	·	●					●		●			●	●	·	●		·	●	·	

£ - unit price multiplier

The unit price multiplier column gives an idea of the relative costs of each option over a 60 year life - not just the initial cost of the frame but also the costs of painting and repainting. (See Painting Cycles on page 123.)

Production Impacts

Production impacts include ratings for environmental impact during extraction, processing, production and distribution phases of a product's life. The issue of maintenance, and the impact of paints and other materials used, is crucial in assessing the full life-cycle impact of a window frame choice. In this Table we have therefore included the impact of paints etc. as a second, separate right-hand 'blob' under each heading. This was thought to give the clearest indication of the overall impact assessment. The ratings for paints etc. were taken from chapter 11, 'Paints and Stains for Joinery'. See that chapter for a detailed explanation of these assessments. For steel and aluminium frames we have assumed that synthetic solvent-based paints will be used. Additional allowance would have to be made if a timber sub-frame were used.

Post Production

The impacts of a window frame choice after installation and during its normal life are assessed as a combination of frame and maintenance.

10.4 Product Analysis

The analysis below gives further details about the criticisms noted on the Product Table. The ratings on the table include those for paints used in the coating and maintenance of window frames as appropriate. These ratings are not included in the analysis below - see chapter 11 for more details.

(a) Aluminium

Product

Aluminium windows may be fixed directly to the wall opening, or, more usually for replacement windows, mounted on timber (often tropical hardwood) sub-frames.[2] The rating on the Product Table is for factory finished aluminium plus a conventional paint maintenance programme beginning after approximately 15 years. Extra allowance should be made for a timber sub-frame if used. Although aluminium can withstand a certain amount of exposure unprotected from the weather, it appears that it will lose its brightness and can become pitted by corrosion.[2] The British Standard for aluminium windows (BS 4873) therefore requires anodising, or finishing with liquid or powder-applied organic coating.

Manufacture

Energy Use

Aluminium has an extremely high embodied energy of 180-240MJ kg^{-1}.[(42)] (or 103,500 Btu at point of use)[43] The aluminium industry accounts for 1.4% of energy consumption worldwide,[42] the principle energy source being electricity. Recycled aluminium gives an 80%-95% energy saving over the virgin resource at 10 to 18 MJ kg^{-1}.[42,43] The production of aluminium uses energy for the heating of initial bauxite-caustic soda solutions, for the drying of precipitates, for the creation of electrodes which are eaten up in the process, and for the final electrolytic reduction process.[3] Finishing processes such as casting or rolling require further energy input. Bear in mind that most embodied energy figures are quoted on a per tonne or per kilogram basis - which ignores aluminium's low density compared to say steel. Of the four aluminium smelters in the UK, though the two small Scottish plants use hydro-power, the larger plants use coal (Lynemouth) and national grid (nuclear) electricity (Anglesey).[4] It is claimed by some commentators that energy consumption figures for aluminium can be misleading, as the principle energy source for virgin aluminium manufacture is electricity produced from hydroelectric plant and is therefore a renewable resource.[42]

Resource Use (bio)

Bauxite strip mining causes some loss of tropical forest.[43] The flooding of valleys to produce hydroelectric power schemes often results in the loss of tropical forest and wildlife habitat, and the uprooting of large numbers of people.

Resource Use (non-bio)

Bauxite, the ore from which aluminium is derived, comprises 8% of the earths crust.[43] At current rates of consumption, this will serve for 600 years supply, although there are only 80 years of economically exploitable reserves with current market conditions.[42]

Global Warming

The electrolytic smelting of aluminium essentially comprises the reaction of aluminia oxide and carbon (from the electrode) to form aluminium and carbon dioxide, the greenhouse gas.[6] Globally this CO_2 production is insignificant compared to the contribution from fossil fuel burning, but compared to iron and steel, aluminium produces twice as much CO_2 per tonne of metal (though allowance should perhaps be made for the lower density of aluminium).

Nitrous oxide emissions are also associated with aluminium production.[7]

One tonne of aluminium produced consumes energy equivalent to 26 to 37 tonnes of CO_2 - but most imported aluminium is produced by hydroelectric power with very low CO_2 emission consequences.[42]

Acid Rain

SO_2 and NOx are released when fossil fuels are burned at all stages of manufacture, to produce electricity (see 'global warming' above) and in gas-fired furnaces.[42]

Photochemical Smog

The Nitrous Oxide emissions associated with aluminium production also contribute to photochemical smogs.[7]

Toxics

Bauxite refining yields large volumes of mud containing

Combinations and Composites

There are a huge variety of combinations of materials and composite window frames available. Some companies supply aluminium framed windows only with Brazilian mahogany sub-frames. Others use galvanised steel inserts in uPVC for extra strength. Becoming more common is aluminium external facing for a timber frame. One composite is actually available with aluminium externally, a wood face internally, and a uPVC thermal break between.[40]

The Product Table would get unwieldy if we tried to include all the possible combinations, so just the basic materials options are shown. We shall leave it up to the reader to try to estimate the combined impact of two or more options if considering using a composite frame.

trace amounts of hazardous materials, including 0.02kg spent 'potliner' (a hazardous waste) for every 1kg aluminium produced.[43]

Fabrication and finishing of aluminium may produce heavy metal sludges and large amounts of waste water requiring treatment to remove toxic chemicals.[43]

Aluminium processes are prescribed for air pollution control in the UK by the Environmental Protection Act 1990,[44] and emissions include hydrogen fluoride, hydrocarbons, nickel, electrode carbon, and volatile organic compounds including isocyanates.[42]

Metal smelting industries are second only to the chemicals industry in terms of total emissions of toxics to the environment.[16]

Aluminium plants in the UK have been frequently criticised for high levels of discharge of toxic heavy metals to sewers.[45]

Emissions of dioxins have also been associated with secondary aluminium smelting.[10]

Other

The open-cast mining of the ore, bauxite, and of the limestone needed for processing can have significant local impact, bauxite mining leaving behind particularly massive spoil heaps.[46]

The association between aluminium smelting and large scale hydro-electric dams in third world countries is well known. So too is the damage such schemes cause to both human communities and to the natural environment.[6,11]

Post Production

Thermal Performance

Even with a 'thermal break' - an insert of plastic or other insulating material - aluminium window frames cause twice as great a heat loss as either timber or plastic.[12]

Recycling/Reuse/Disposal

Aluminium is normally easily recycled, saving vast amounts of energy compared to making new, but powder coated aluminium is not recyclable.[13] Anodised aluminium would appear therefore to be better for recycling at the end of its useful life. However, powder coating is usually necessary because aluminium's natural corrosion resistance cannot cope with the acidity of most rainwater in the UK. Water coming through cementitious materials can also seriously attack aluminium.

Health Hazards

The main potential health hazard associated with aluminium as with other painted window frames is dust or fume inhalation by painters whilst sanding or burning off paint in the maintenance phase.[14]

(b) Steel

Product

Older steel windows (pre-1950s) have suffered from rust problems, with rust even forming on the inside from condensation moisture.[15] Modern steel windows are now usually made from standard rolled steel sections and galvanised for corrosion protection. As with aluminium windows, they may or may not be on a hardwood timber sub-frame.[2]

Production

The impacts of steel production are detailed in Chapter 13, 'Rainwater Goods', p.174.

Post Production

Thermal Performance

Even with a 'thermal break' - an insert of plastic or other insulating material - steel window frames cause nearly twice as great a heat loss as either timber or plastic.[12]

Recyclability

As with aluminium windows, paint coatings on steel can interfere with its ability to be easily and cleanly recycled.

Health Hazards

The main potential health hazard associated with steel as with all painted window frames is dust or fume inhalation by painters whilst sanding or burning off paint in the maintenance phase.[14]

(c) Timber

The following analysis covers all timber options listed on the table. Where distinctions can be drawn between the different timber types, these differences are mentioned below. See chapter 7 (Timber) for more details, especially of the Forest Stewardship Council's certification scheme for well-managed forestry.

Product

Softwoods are the traditional timber for windows in the UK. Most are made from preservative treated European redwood (a temperate softwood), but hardwoods are also used in significant quantities.[2] Preservative treatment is usually by the double vacuum process with light organic solvents.

It is generally believed that wood for windows must be either from a naturally durable species, or treated with preservative. Even with naturally durable species such as some tropical hardwoods, it is increasingly common for sapwood (which is always perishable) to be included, unless deliberately excluded by the specifier. Thus preservative treatments are commonly specified for hardwoods too.[2]

Temperate hardwoods such as oak and chestnut also offer a durable option, although other temperate hardwoods are probably non-durable and require treatment. Again, sapwood is not always excluded. Sometimes oak may be left untreated. It goes a nice silvery grey when weathered, but there is some risk of surface checking in very exposed situations.[21]

Factory finished wood windows are becoming more common. The surface coating will usually last around 10 years before refurbishment is needed, twice as long as site-painted windows. Thereafter repainting will presumably be at the usual intervals, about every five years. Factory finished frames will however be coated all round, so hidden surfaces which don't normally get anything more than a first coat of primer will be much better protected. Durability should therefore be better.

Production

Embodied Energy
Transport energy may be significant for imported timber, especially hardwoods from the Pacific countries such as Papua New Guinea or Indonesia (see chapter 7), but is much less significant than the embodied energy of other window frame materials. For all types of timber, processing, and possibly kiln drying also take energy, but these are again probably relatively insignificant.

Resources (bio)
Although timber is often seen as a 'renewable' resource, modern forestry practices are not always 'sustainable' or managed in the interests of long-term viability. Much tropical hardwood, and even temperate hard and softwoods are cut from old-growth forests, and even if replanting does take place, a natural biological capital resource is being replaced by an agribusiness system of doubtful sustainability. Timber that is certified by an FSC accredited agency as coming from a well managed source is an exception to this.

Global Warming
Tropical timber production is, in its present unregulated form, is causing loss of forest. It also opens up the forests to further destruction by other processes (settlement, mining & industry). Tropical deforestation from all causes is responsible for a large proportion (18%) of global warming.[22] Conversion of old growth forest to plantation may also cause increases in greenhouse gas emissions.[23] Responsible forestry which ensures replanting over the long-term will make no net contribution to global warming, and may cause a net decrease.

Toxics
Preservative pre-treatment is the norm now for timber used for window frames. Even supposedly durable tropical hardwoods such as meranti and luan often 'need' treatment, as sapwood is sometimes not excluded.[21]
Preservative treatments for timber are covered more fully in a chapter 9.
Plantation-grown timber may well have been the subject of toxic pesticide treatments - seedlings are regularly treated with gamma-HCH (lindane), and aerial spraying of forests is sometime resorted to in order to control pests.[24]

Other
Human rights violations may occur in both tropical and temperate regions[25,26] as indigenous peoples are removed from their traditional forest homes.

Post Production

Thermal Performance
Timber window frames have about the same insulation value as double glazing, so although better than metal frames, they still contribute to heat losses.[12]

Painting Cycles

The following table shows the time in years to the first and subsequent repainting to be expected for different frame types and coatings. (2,15,21)

Frame Type	First Repainting (yrs)	Subsequent Repainting (yrs)
Aluminium - powder coated	15	5
Steel - Powder Coated	15	3-7
Steel - Painted	5-7	3-7
Wood - Stained	3	3
Wood - Site Painted	5	5
Wood - Factory-Painted	10+	5

Recyclability
The existence of both reclaimed window frames in salvage yards, and suppliers of window frames made from reclaimed timber, attests to the recyclability of timber and joinery, although most old wooden windows are unlikely to be recycled as such.

Health Hazards
The main potential health hazard associated with timber window frames is dust or fume inhalation by painters whilst sanding or burning off paint in the maintenance phase.[14]
In a house fire, timber windows will burn, but the contribution to poisonous fumes from a window will be insignificant compared to fumes from other timber sources in the building.[27]

(d) uPVC

Product

Despite being around for 30 years, uPVC is still a relatively new product, and formulations are being developed continuously. It is the most common material for replacement windows, but has not made such an impact for new buildings. The Building Research Establishment considers that, although uPVC windows have not been in use long enough to assess their durability in the long term, "experience to date is generally encouraging."[2]
At one time uPVC extrusions were much bulkier than timber, and were therefore considered ugly by many. Improvements in materials and design have meant smaller and more stable profiles are now available.[12]
Nevertheless, uPVC is not as stiff as wood or metal, and therefore, in all but the shortest sections, reinforcement is often used. This is usually made from galvanised steel tube, or sometimes aluminium. (The use of such reinforcement has not been included in the ratings here,

and reinforced uPVC windows should be considered as a composite product - see 'Combinations and Composites' Box, p.125)

Plain white is the most common finish, but many windows are now produced in colours or printed-foil patterns (wood grain etc.).[27]

Production

Details the impact of uPVC production can be found in chapter 13 (rainwater goods)

Post Production

The impacts below relate specifically to uPVC windows. General post production impacts of uPVC can be found in chapter 13 (Rainwater Goods).

Thermal Performance

uPVC window frames have about the same insulation value as for wood and for double glazing, so although not as bad as metal frames, they still contribute to heat losses.[12]

Recyclability

Cycles of processing and use degrade the PVC polymer, and because window profiles need high-grade polymer for durability, recycled content is strictly limited. Current UK standards allow only 10% of reprocessed material (originating within manufacturing process). Newer technology allows a core of recycled uPVC (from used windows) to be made with an outer layer of new polymer. There are only draft standards covering this at present, which include requirement for adding extra impact additives and stabilisers.[27]

Because of the many different additives present in PVC, it is "impossible to recycle in the true sense of the word". If disposed of in landfill, leaching of toxic additives such as plasticisers and heavy metal stabilisers is possible. The chlorine content of PVC makes it "completely unsuitable for incineration" due to the creation of dioxins and other organochlorines as well as highly corrosive hydrogen chloride gas. And in fact burning 1 tonne of PVC creates 0.9 tonnes of waste salts which are still toxic and need disposing of.[28]

Health Hazards

Tests comparing the effects in building fires of wood and uPVC window frames, performed by the Fire Research Station and the British Plastics Federation, concluded that uPVC windows created no new hazards compared to wood. Although uPVC gives off poisonous hydrogen chloride vapours when burnt, the quantity of these is such that CO fumes from the fire in the room itself are a more serious, immediate hazard.[27] However, the research did not account for the longer term effects of toxins given off by burning PVC.

Waste offcuts or old PVC should never be burnt on bonfires - phosgene, dioxins and hydrogen chloride fumes given off are all extremely dangerous. All new plastic products have that characteristic 'plasticy' smell - this is caused by off-gassing of the many different constituent chemicals. PVC may release the highly carcinogenic vinyl chloride monomer for some time after manufacture.[33]

Alert

The environmental group Greenpeace is campaigning world-wide for an end to all major industrial chlorine chemistry, including the manufacture of PVC.[28] See Toxics above and also chapter 13 - *Rainwater Goods* which also includes a response to some of these criticisms by the industry body, the British Plastics Federation.

Durability, Expectations and Aesthetics

The effects of the weather on uPVC include: loss of surface gloss; 'chalking'; reduction of impact resistance; yellowing. Proper formulation can control these effects within acceptable limits. Painting of uPVC is technically difficult, and not possible with normal paints.[27]

Unless regularly washed down, lower rails in particular are susceptible to dirt retention and subsequent discoloration. Performance and colour-fastness of coloured, surface coated and foil covered materials is less well proven than for white.[2]

If uPVC windows suffer physical damage it would seem that they are much more difficult to repair satisfactorily than timber - especially if the extruded section is no longer manufactured.

The aesthetic aspect of maintenance cycles is an area where there has been little comment. uPVC windows start off bright and white. We have very high expectations of them as plastics, and would like them not to change over their whole life. In reality their gloss might fade and their whiteness yellows. Many other building materials are generally thought to become more beautiful as they age, but for some reason this does not seem to apply to plastics. One explanation might be that the 'no maintenance' regime for plastics means that, once faded, it will remain faded for the rest of its life. Other materials that need regular repainting will, in contrast, have a regular 'face lift' every five years or so. Perhaps as the number of ageing uPVC windows increases, manufacturers will come up with paints or other maintenance systems for uPVC that will help restore their appearance. Of course, once this happens, uPVC will have lost much of its 'no maintenance' advantage over timber.

10.5 Design

This section looks at what can be done at the specification and design stage to ensure the long-term durability of wood windows. See references 2, 3, 21, 34, 35, 36 and 37 for more details.

The following points should be emphasised when choosing, specifying or designing timber windows:

10.5.1 Detailing

- Are the end grains of timbers within joints properly sealed against moisture? These are one of the most vulnerable points in external joinery. Sealing end grains may need to be done before assembly.

- Does the design rapidly shed rain water? There should be no water traps, and no horizontal surfaces, especially at sills and the bottom of sashes etc. (A minimum slope of 5-7° is recommended.)

- There should be no sharp arises (edges of profiles). Rounded arises are much better for maintaining paint thickness and adhesion, and they aid rapid drainage too.

- Are weather seals located at the back of the casement, away from wet part? Water-logging can seriously affect the life of seals.

- Glazing channels should be ventilated and drained to preserve double-glazing sealed units. (The seals of sealed units will deteriorate if wet.) Drained systems are favoured because of the difficulty of ensuring, especially at the bottom edge, that absolutely no tiny cracks are formed between glass and glazing bar or sealant.

- Air pressure relief spaces behind the outer face of each joint should be standard nowadays. They ensure that wind pressure doesn't force moisture through weatherseals etc.

- Screws, dowels etc. in the external face should be avoided due to capillary risk.

The Beautiful North

Windows placed on the north side of a building, made small to save energy, can still have the best views. Being small need not diminish the view if there is careful placement and orientation from the viewers angle. And views are most beautifully illuminated with light from south. [39]

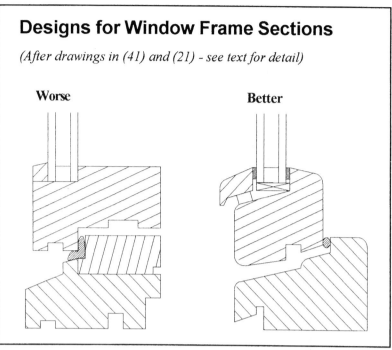

Designs for Window Frame Sections

(After drawings in (41) and (21) - see text for detail)

Worse   **Better**

10.5.2 Installation

- There are two, possibly three, conflicting requirements for the position in the wall within which a window is fixed. For the sake of durability, the further a window is set back from the outer face of a wall, nearer the warm, dry zone of the inner leaf, the better protected from the elements it will be. Better thermal insulation will be achieved, however, with the window in the more conventional position (for the UK) in the outer leaf of the wall (so long as thermal bridging is avoided). This conventional position is preferred by builders because of tradition, and also because it is easier not to close the cavity with a facing material. It also allows for a wide internal window sill.
  The most practical solution is to place the window midway between these extremes, possibly using the window subframe to close the cavity. Flange fitting of frames into subframes also makes draught sealing etc. easier.[21,38]

- Using the window frame as a former whilst building the wall risks damage to painted surfaces. Special resizable, reusable formers are available.

- If the window frame is not factory finished, fully painting what will be invisible surfaces once installed, will enhance durability further.

Positioning of window frame in cavity wall

Typical Position
Better thermal performance & wide inner sill but poor protection & durability for window

Best Protection
Window sheltered but poorer thermal performance and no inner sill

Best Compromise?
plus traditional Scottish detailing - outside leaf covers joint

inside

inside

inside

outside

outside

outside

Factory Finishes for Timber

Although factory-applied finishes might last twice as long as site-applied paint before the first refurbishment is required (perhaps 10+ as opposed to 5 years), if they thereafter need painting every five years like conventionally-treated windows, then the overall impact of the paint system over a life of sixty or a hundred years is going to be little different. However, factory-finished windows will usually be fully coated all round, including backs and ends, parts not normally covered by anything but primer on site-finished windows. The overall durability and longevity of the frame is thus likely to be enhanced. Factory-finishing is becoming more common, although site-finishing is still the norm in the UK.

"Several smaller windows are better than one large one, not only because, from the energy saving point of view, for the same heat loss there is a better distribution of light, avoiding quantitative extremes, but also for quality. The light is more full of health-giving - and aesthetically satisfying - life. Also you get two views instead of one, which helps you to orientate yourself."

Christopher Day (1)

Shutters and Curtains

Even with double-glazed, low-emissivity windows, heat loss from windows can be considerable. Internal insulating shutters can make a tremendous difference. Even poorly fitting shutters could save 40% on fuel bills.[39] Good fit, especially around bottom and sides is important in preventing cold air trapped between shutter and glass from re-entering the room. All sorts of materials could be used, from plastic foam panels encased in timber frame and thin ply, to roll-down inflatable or expanding blinds Some options are even translucent and therefore let in at least some light so they can be used on cold winter days.

Heavy curtains can have a remarkable insluating effect too, though they may cause uncomfortable pools of cooler air at floor level. Surprisngly, the best curtains may be net curtains, which by restricting the air-flow across the face of a window, can insulate well without causing cold feet.[38]

10.6 Suppliers

J.R.Nelson, Dupree Partnership Ltd, The Sawmill, Will's Farm, Newchurch, Romney Marsh, Kent, TN29 0DT (tel: 01233 733361)

Contact: Jeremy Nelson

Reclaimed pitch pine windows and doors to standard or specification.

Mumford & Wood Ltd, Hallsford Bridge Ind. Estate, Ongar, Essex, CM5 9RB (tel: 01277 362401 fax: 01277 365093)

Wide range of styles for refurbishment or new work.

Norwegian Building Agency, Nor-Dan, 31 Windsor Drive, Tuffley, Gloucester, Gloucestershire, GL4 0QJ (tel: 01452 311379 fax: 01452 311402)

Contact: Colin Collins

According to the company, a typical house with Nor-Dan Super-insulated Windows can save 4,250KWH compared to glazing U-Values of 2.9 W/m2k. Or 3,000KWH compared to glazing U-Values of 1.8 W/m2K. Nor-Dan claims to take very serious consideration of environmental issues.

Nor-Dan's Super-insulated Window won the Swedish State's Energy Works (NUTEK) 1992 contest for window manufacturers from several countries to produce a better insulated window. The criteria included not only a low U-value (less than 0.9 W/m2K), but also good light transmission, being able to withstand high wind loadings and achieve high levels of water tightness, ease of use and maintenance, as well as visual aesthetics and commercial viability.

Nor-Dan windows featured in the super insulated bungalows at the Milton Keynes Energy World Exhibition.

Super-insulated Windows are available as ND Super 2+1 (U Value 0.98 W/m2K) or ND Super 3+1 (0.86 W/m2K), the former being double/triple glazed, and the latter triple/quadruple glazed.

Trayd Fonster Ab, 13 Ashlyn Road, West Meadows Industrial Estate, Derby, Derbyshire, DE21 6XE (tel: 01332 349161 fax: 01332 291119)

Contact: Peter Howe

Double or triple glazed Swedish windows and doors, pre hung with option of prefinished.

Peter King, Conservation Rooflight, The Old Stables, Oxleaze Farm, Filkins, Lechlade, Gloucestershire, GL7 3RB (tel/fax: 01367 85313)

Contact: Peter King

Rooflights, fabricated from steel to replicate traditional cast iron rooflights. Available double glazed.

Centrum Windows (UK),

Unit 12, Murdock Rd, Bicester Oxfordshire, OX6 7PP (tel: 01869 248181 fax: 01869 249693)

High performance windows and doors in seasoned pine. Factory fitted double or triple glazing, weather seals and pre-hung doors. Multipoint locking as standard and a micro-porous staining service available. Made to measure for new or refurbishment contracts

Swedish Window Co Ltd, Earls Colne, Colchester Essex, CO6 2NS England (tel: 0787 223931 fax: 0787 224400)

Contact: Sean Milbank, Director

Imported high performance windows from Scandinavia. Made from Scandinavian Redwood with a choice of double or triple glazing and espagnolette multi-point locking as standard.

Tanums Fonster Ab, 149 Hassock Lane South, Shipley Heanor, Derbyshire, DE75 7JE (tel: 01509 321013)

Swedish, quality high performance double or triple glazed Redwood windows and French doors.

The Velux Co Ltd, Woodside Way, Glenrothes East Fife, KY7 4ND (tel: 01592 772211 fax: 01592 771839)

Woodland Windows, 28 Beck Lane, Macclesfield, Cheshire, SK10 3EX (tel: 04213 78732 (mobile)/ 0161 480 0363)

Contact: David Eadie

Low maintainance windows made from homegrown timbers from managed forests. Finishes are solvent free.

Environmental Construction Products

Eco-Plus Timber Windows, 26 Millmoor Road, Meltham, Huddersfield, West Yorkshire, HD7 3JY. (tel: 01484 845898 fax: 01484 845899)

Environmental Construction Products claim to have extensively researched the environmental impact of all components of window systems. Ecoplus are manufactured in the UK to BS1186 pt1 and 644 pt1, and according to the company, incorporate advanced design detailing. Slow-grown Scandinavian redwood or durable homegrown softwoods are available, whilst preservative pretreatment, where required, uses the Eco inorganic borate system. Ecoplus windows are available in a range of styles: Stormproof and Scandinavian, with High Security and Superwarm options. Natural finishes are available, and glazing is with 'iplus neutral R' sealed units, made to BS5750, offering U-values from 1.1 to 0.4 W/m²K.

N.B. Claims for a product's sustainability in this listing are the suppliers' own, as supplied to the GreenPro database.

10.7 References

1. Timber Trades Journal, 18 March 1995
2. Selecting windows by performance, BRE Digest 377 (Building Research Establishment, Garston, 1992)
3. Construction Materials Reference Book (ed. D K Doran, Butterworth-Heinemann Ltd, Oxford, 1992)
4. UK Minerals Yearbook 1989 (British Geological Survey, Keyworth; Nottingham 1990)
5. Minerals Handbook 1992-93 (Philip Crowson, Stockton Press, New York 1992)
6. Greener Building Products & Services Directory (K Hall & P Warm, Association for Environment Conscious Building, Coaley, Glos.)
7. Metal Industry Sector IPR 2 (Her Majesty's Inspectorate of Pollution, HMSO, London, 1991)
8. The World Environment 1972-1992 - Two Decades of Challenge (eds. M K Tolba et al, UNEP/Chapman & Hall Ltd, London, 1992)
9. The Secret Polluters (Friends of the Earth, UK, July 1992)
10. ENDS Report January 1995 Issue No. 240 (Environmental Data Services)
11. Environmental News Digest vol.10 no.1 (Jan. '92, Friends of the Earth Malaysia)
12. A Practical Guide to Windows - A Building Trades Journal Book (P. Roper, International Thomson Business Publishing, Southwick 1986)
13. The Green Construction Handbook - A Manual for Clients and Construction Professionals (J T Design Build, Bristol 1993)
14. Buildings & Health - The Rosehaugh Guide to the Design, Construction, Use & Management of Buildings (Curwell, March & Venables, RIBA Publications, 1990)
15. The Building Design Easibrief 1993 Edition (Henry Haverstock, Morgan-Grampian (Construction Press) Ltd, London 1993)
16. Environmental Chemistry 2nd Edition (Peter O'Neill, Chapman & Hall, London, 1993)
17. Eco-Renovation - the ecological home-improvement guide (E Harland, Green Books Ltd, Dartington 1993)
18. ENDS Report March 1994 Issue No. 230 Page No. 44 (Environmental Data Services)
19. Chlorine-Free Vol.3 No.1 (Greenpeace International 1994)
20. Dioxins in the Environment, Pollution Paper 27 (Department of the Environment, HMSO, London, 1989)
21. Wood Windows - design, selection and installation (P J Hislop, TRADA Timber Research & Development Association, High Wycombe 1993)
22. Global Warming - the Greenpeace Report (Ed. J Leggett, Oxford University Press, Oxford 1990)
23. Timber: the UK Forestry Industry's 'Think Wood' and 'Forests Forever' campaigns (Friends of the Earth, London)
24. Forestry Practice (ed. B G Hibberd, Forestry Commission, London 1991)
25. Why Forests Are Important (WWF Fact File, World Wide Fund for Nature UK, Godalming 1991)
26. Timber: Types & Sources Briefing Sheet (Friends of the Earth, London 1993)
27. PVC-U windows - BRE Digest 404 (Building Research Establishment, Garston 1995)
28. PVC - Toxic Waste in Disguise (Greenpeace International, Amsterdam 1992)
29. Oil & Energy Trends Annual Statistical Review (Energy Economics Research, Reading 1990)
30. Achieving Zero Dioxin - An Emergency Strategy for Dioxin Elimination (Greenpeace International, London 1994)
31. Review of Accidents Involving Chlorine (G. Drogaris, Community Documentation Centre on Industrial Risk, Commission of the European Communities, Luxembourg 1992)
32. Production & Polymerisation of Organic Monomers IPR 4/6 (Her Majesty's Inspectorate of Pollution, HMSO, London 1993)
33. C for Chemicals (M Birkin & B Price, Green Print, London 1989)
34. Double glazing for heat and sound insulation - BRE Digest 379 (Building Research Establishment, Garston 1993)
35. Avoiding joinery decay by design - IP 10/80 (J Carey, Building Research Establishment, Garston 1980)
36. Preventing decay in external joinery - BRE Digest 304 (Building Research Establishment, garston 1985)
37. Energy Efficient Building: A Design Guide (Susan Roaf and Mary Hancock, Blackwell Scientific, Oxford 1992)
38. Heat losses through windows - IP 12/93 (R Rayment, P Fishwick, P Rose & M Seymour, Building Research Establishment, Garston 1993)
39. Gentle Architecture (Malcolm Wells, McGraw-Hill, New York 1982)
40. Architects Journal focus June 95 (Emap Business Communications, London 1995)
41. The Good Wood Manual - specifying alternatives to non-renewable tropical hardwoods (Ed. Ian Grant, Friends of the Earth, London)
42. The Environmental Impact of Building and Construction Materials, Volume C: Metals. (N. Howard). CIRIA, June 1995.
43. The Greening of the Whitehouse. http://Solstice.crest.org/environment/gotwh/general/materials.html
44. Environmental Protection Act 1990 Pt.1. Proceses prescribed for air pollution control by Local Authorities, Secretary of State's Guidance - Aluminium and aluminium alloy processes. PG2/6(91). (Department of the Environment). HMSO 1991
45. The Secret Polluters (Friends of the Earth UK). July 1992
46. Ecological Building Factpack. (Peacock and Gaylard). Tangent Design Books, Leicester. 1992

"The essence of green architecture is not that there is a single correct answer in any situation, but that designers should be aware of the factors to be evaluated when materials and components are specified."

(Brenda & Robert Vale, 'Green Architecture')

Paints and Stains for Joinery   11

11.1 Scope of this Chapter

This chapter looks at the main options available in the market for paints, stains and varnishes, which perform both decorative and protective functions for joinery. Paints, stains and varnishes are made in similar ways, their particular characteristics being chosen by the precise formulation. In this report we have drawn distinctions between products by their basic solvent and the source of their main ingredients - by how they are made, and not by type (paint, stain or varnish). Throughout the chapter, when we refer to 'paints' we include paints, stains and varnishes unless stated otherwise.

11.5 Environment Conscious Suppliers

Auro Organic Paints Unit 1, Goldstones Farm, Ashdown, Saffron Walden, Essex CB10 2LZ (tel: 01799 584888)

Auro Organic Paints are made in Germany with plant-derived oils and solvents. Auro claims a "total commitment to sustainable ecology". Full contents listings are provided. Products for coating wood include stains, gloss paints and preservatives, available in a variety of natural colours. All use plant-derived oil solvents such as citrus peel oil and turpentine. Auro Organic Paints claim their exterior gloss system is "expected to last 8 to 10 years if properly applied on sound surfaces in normal conditions".

Biofa Natural Paints 5 School Road, Kidlington, Oxford OX5 2HB

Biofa paints are also plant-based and manufactured in Germany, have full ingredients listings, and include paints, varnishes and stains. Their range includes a water-based wood stain for interior use.

Environmental Paints Ltd Unit 11, Dunscar Industrial Estate, Blackburn Road, Egerton, Bolton BL7 9PQ

Environmental Paints produce the ECOS (Environment Conscious Odourless Solvent-free) range of paints, which includes gloss as well as emulsions. These paints are not advertised as being plant-based or organic, so we must assume they are water-based synthetics.

Nutshell Natural Paints, Hamlyn House, Mardale Way, Buckfastleigh, Devon TQ11 0NR (tel: 01364 642892 fax: 01364 643888)

Nutshell products include an interior/exterior stain/varnish and an interior wood wax, made, "using only proven natural raw materials such as harvested tree resins and oils, chalk, well-water, mineral pigments, beeswax and talcum." Their linseed oil is organically grown, and they claim not use any petrochemical substances at all. The company will take back empty paint containers to be returned to the manufacturer for re-use.

Ostermann & Scheiwe Osmo House, 26 Swakeleys Drive, Oxbridge, Middlesex UB10 9DX (tel: 01895 252171)

Ostermann & Scheiwe produce a range of interior and exterior wood finishes based on natural oils and waxes containing no biocides or preservatives. Medical grade white spirit is used as the solvent, which contains none of the usual impurities (benzene etc.).

D.I.Y. ING? EXPAND YOUR PSYCHIC LANDSCAPE

USE SOLVENT BASED PAINTS

11.6 References

1. ENDS Report 240, January 1995 (Environmental Data Services)
2. Dictionary of Environmental Science & Technology (Andrew Porteus, John Wiley & Sons, Chichester, 1992)
3. The World Environment 1972-1992 - Two Decades of Challenge (M. K. Tolba & O. A. El-Kholy (eds), Chapman & Hall, London, for The United Nations Environment Programme, 1992)
4. Green Design: A Guide to the Environmental Impact of Building Materials (Avril Fox & Robin Murrell, Architecture Design & Technology Press, London, 1989)
5. Greener Building: Products & Services Directory (Keith Hall & Peter Warm, Association for Environment Conscious Building, Coaley)
6. Hazardous Building Materials (S R Curwell & C G Mach, E & F N Spon Ltd, 1986)
7. Out of the Woods - Environmental Timber Frame Design for Self Build (Pat Borer & Cindy Harris, Centre for Alternative Technology Publications, Machynlleth, 1994)
8. Buildings & Health - The Rosehaugh Guide to the Design, Construction, Use & Management of Buildings (Curwell, March & Venables, RIBA Publications, 1990)
9. Eco-Renovation: The ecological home improvement guide (Edward Harland, Green Books, Dartington, 1993)
10. Pocket Card for Professional Painters & Decorators (Paintmakers Association, UCATT and NFPDC, 1991)
11. Painting exterior wood, Digest 354 (Building Research Establishment, Garston, 1990)
12. Water-borne coatings for exterior wood, IP 4/94 (E R Miller & J Boxall, Building Research Establishment, Garston, 1994)
13. Natural finishes for exterior wood, Digest 387 (Building Research Establishment, Garston, 1993)
14. Factory-applied stain basecoats for exterior joinery, IP 2/92 (J Boxall, Building Research Establishment, Garston, 1992)
15. The UK Environment (Department of the Environment, HMSO, London, 1992)
16. Which? magazine, April 1994 (Consumers' Association, London)
17. Oil & Energy Trends Annual Statistical Review (, Energy Economics Research, Reading, 1990)
18. C for Chemicals - Chemical Hazards and How to Avoid Them (M. Birkin & B. Price, Green Print, 1989)
19. ENDS Report 241, February 1995 (Environmental Data Services)
20. Auro Paints Brochure 1995
21. 'The Paint Industry Today': Report of an address by Dr. Fischer to the AECB on 22/3/93
22. Preventing Decay in External Joinary. BRE Digest 304. Building Research Establishment, Gaston 1985
23. External Joinary: Endgrain Sealers and Moisture Control. BRE Information Paper IP 20/87. Building Research Establishment, Garston, Dec. 1987
24. Ethical Consumer Magazine Issue 30, June/July 1994
25. Blueprint for a Green Planet (Seymour & Girardet). Doring Kindersley 1987.

> *"The present rate of paint use is unsustainable in many different ways, largely because of the many toxic chemicals involved. We need to look carefully at all the possible alternatives to the toxic mixtures that are sold as paint."*
>
> Edward Harland - Eco-Renovation[9]

Roofing Materials 12

12.1 Scope of this Chapter

This chapter looks at the environmental impacts of the main options available in the market for roof coverings. These include clay tile, natural slate, concrete tile, fibre-cement tile, polymer modified cement tile, reconstructed slate, asphalt shingles, steel, stainless steel, aluminium, lead and copper sheet. Wood shingles are covered in the Alternatives section. Concrete and asphalt/felt sheet are not covered per se, but their impacts can be considered as virtually the same as the corresponding tile product. Gravel and hot bitumen systems, roofing underlay, sarking and flashing detail are not covered in this report. Specific multi-component systems are not covered, although the component parts found in most systems are discussed individually. The insulation value of each material is not covered in this report, but roofing insulation materials are covered in chapter 5.

12.2 Introduction

Of all the components which make up a building, roofing plays an especially critical role, performing important functions of insulation from heat and cold, protection from rain and wind, and provision of shade.[36] Roofing must withstand extreme conditions - strong winds, temperature swings, long term exposure to ultra-violet light and extreme precipitation - the precise criteria depending on the climate in which it is to be used.

Durability is of prime importance from both an economic and environmental viewpoint. A long lifespan product reduces manufacturing impacts and contributes less to the waste disposal burden, as the roofing will require replacement less often and may be reused after the lifetime of the building has elapsed.

However, there is often a trade-off between increased durability and increased environmental impact during manufacture - for example, higher quality, high durability clay tiles generally require a longer firing time than poorer quality clay tiles[1] and thus consume more energy in manufacture. Also, for short lifespan buildings, it may be considered more environmentally sound to use a short lifespan product with low production impact, particularly if reuse of the roofing material is unlikely.

Maintenance should also be considered, particularly the replacement of damaged tiles. For example, in Milton Keynes, the concrete tiles for many of the houses were made in 'one-off' batches; when tiles were lost during the storms of October 1987, these could not be replaced with matching tiles and many homeowners opted to re-tile the entire roof rather than have a colour mis-match.[58]

There is also potential for using roofs as more than simply a shelter from the elements, by utilising rainwater catching devices for irrigation or potable water (discussed in chapter 13), planted roofs to create green space or photovoltaic systems for energy production, which are discussed in the Alternatives section, page 164.

12.2.1 Fixings, Flashing and Durability

Roofing systems are only as good as the weakest link and so fixings and flashing details should be designed with the lifetime of the primary roofcovering material in mind, taking into account any future maintenance.

Slates and tiles are traditionally fixed to wooden battens using iron, steel or even wooden pegs. These eventually deteriorate, allowing the tile to fall from the roof,[3] limiting its lifespan. Galvanised nails are only a slight improvement, as the zinc coating is easily scratched, allowing corrosion of the steel beneath. Stainless steel or aluminium nails are recommended by BS 5534: Pt 2 1986,[3] and copper and

Packaging

Packaging of tiles is usually timber pallets with shrink wrap polyethylene - usually 10 plain or 32 interlocking tiles per pack.[33] The environmental impacts of packaging can be reduced if the supplier has a system for collecting and reusing pallets.

silicon bronze nails also have good corrosion resistence.[33] The roof may also require additional fixings to protect against wind, the most effective method reported to be the use of aluminium clips.[3]

Fixings were traditionally combined with sand/cement bedding at the ridge, but dry systems have been developed which obviate any requirement for this. These systems are easily installed, incorporate ventilation apertures which overcome condensation in insulated roofspaces, and make frost and wind damage negligible - thus making the roof almost maintenance free.[33] This is also likely to extend the lifespan of a roof, which is an environmental advantage.

12.2.2 Roofing Materials

(a) Clay Tiles

Raw clay in the UK is generally kaolinite with crystalline aluminium oxide and traces of quartz, mica and iron oxide. Binding is achieved by vitrification of the minerals (heating to form a glass). The longer the firing time, the greater the density, strength, hardness, chemical and frost resistance and dimensional stability of the clay product.[33] However, long firing times also increase the brittleness, and the embodied energy of a tile which is already fairly high due to high firing temperatures required (generally around 1100°C.)[33] Handmade tiles tend to have greater strength and more uniform consistency, making them less susceptible to breakage during handling, and less vulnerable to frost damage than machine manufactured tiles.[33]

Reclaimed clay tiles are widely available and present an environmentally preferable alternative to new tiles (see p.168 for details of SALVO, the reclamation organisation).

(b) Slate

Slate is the 'traditional' roofing material, available in a variety of colours; grey to black, red, blue, purple, green and brown. The chief source of slate in the UK is North Wales. This is typically blue, purple, blue-grey, green and dark grey.[31] Cornish slate is grey and grey-green or sometimes red, and is somewhat thicker than Welsh slate. Slate from the Lake District is generally green or blue; Calcium carbonate in the green slates is attacked by sulphur gases, but this is considered less important due to the thickness of Lake District slates.[31]

Originally slate manufacture was a locally based industry, with slates obtained from local quarries. Due to reduced production costs facilitated by cheaper labour, the industry is now global with slates being imported from Europe, South America, Asia and Africa.[31, 58] This will add to the energy use due to long distance transportation - although this may be offset somewhat by less mechanised, more labour intensive quarrying methods - and may lead to greater local environmental degradation due to the less strict environmental legislation in third world countries.

In 1993, 26,238,000 tonnes of roofing slate were delivered in the UK.[37]

Reclaimed slates are widely available, and present an environmentally preferable alternative to new slates (see p.168).

(c) Concrete/Cement Tiles & Slates

'Artificial' slates and tiles can be mass produced cheaply and to high quality standards and provide an attractive and durable, aesthetic and cost effective alternative to 'natural' roofing materials. They are currently distinguishable from natural tiles, but manufacturers claim that future developments are likely to make them indistinguishable.[33] One potential downside to this is that concrete tiles can imitate natural 'weathered' tile, with the advantages of quality assurance and cost savings. This may displace the market for the more environmentally benign reclaimed tiles.

However, light weight concrete tiles are advantageous in that they can be used to replace tiles on older roofs without the cost and materials use of roof strengthening. Improvements in materials technology have allowed tiles to be made thinner and stronger, with resulting savings in materials intensity.[33]

Concrete tiles and slates are manufactured by extrusion, from portland cement, sand and water, with inert inorganic pigments.[33, 37] Fibre-cement tiles are manufactured using portland cement with a fibre matrix, which improves strength and gives a slate like texture. Asbestos fibres were traditionally used, but due to health concerns these were replaced in the 1980s by blends of glass fibre, cellulose and resin based compounds,[33] including polypropylene, polyvinyl alcohol, polyacrylonitrile, sisal and coir and glass fibre.[47] Most substitutes are safer to manufacture than asbestos because the reinforcing fibres are of wider diameter, and therefore less respirable. However, reinforcement is generally less efficient than with asbestos due to the wider fibre diameter, therefore new products are less strong. Cellulose fibres are less heat resistant than asbestos, but have the environmental advantage of being sourced from renewable resources. Long term durability is difficult to assess in new products, but despite problems with alternatives developed in the early 'eighties, products developed within the last eight years seem to be performing satisfactorily.[47]

Details of safety procedures for the removal of existing asbestos fibre-cement tiles are detailed in the DoE book 'Asbestos Materials in Buildings'[47] and the leaflet 'Asbestos in Housing'.[48]

In developing countries, where industrial products such as glass fibre and polymer fibres are less suited to the socioeconomic circumstances, plant based fibres are commonly used in cement tiles. The most common fibres are sisal and coir.[36]

Fibre reinforced cement tiles using sisal and coir without any additional manufactured fibres have a shorter life expectancy than those made with 'industrial' fibres, and can be expected to last between 10 and 15 years.[36] 25,606,000m² concrete roof tiles were delivered in the UK in 1993.[37]

(d) Asphalt Tile

Asphalt tiles are a popular roofing material in the USA, where they take up a 75% share of the roofing materials market. Asphalt shingles are less popular in the UK, although asphalt-felt sheeting is popular for flat roofing. Asphalt shingles consist of a fibrous mat with an asphalt coating. Traditionally, wood, paper or other organic fibres were used, but since the 1980s, these have been largely replaced in the market by fibreglass.[1] This switch from a mainly recycled, renewable resource (felt) to a highly manufactured material (fibreglass) is something of a backwards step environmentally, although fibreglass mat shingles reportedly require significantly less asphalt in their manufacture.[1]

Due to low profit margins at the low-cost end of the roofing tile market forcing manufacturing standards down, economy asphalt shingles reportedly suffer problems with durability,[1] and are therefore not recommended as an environmental option.

(e) Sheet Metal

Materials on the market range from flat sheets of coated steel formed on site, to preformed sheets, and individual steel or aluminium shingles.[1] The market is dominated by profiled steel panels.

Sheet metal roofing is commonly used for industrial units, agricultural buildings and lightweight structures such as garages and shelters. The main exception to this is copper roofing, which weathers to an attractive green colour, and is used as roofing for homes and 'prestige' buildings

The most common material is steel, coated with zinc or aluminium, or a 45:55 mixture of zinc and aluminium (Galvalume). This is generally covered with an organic coating - usually PVF2 (Polyvinylidene fluoride), polyester, siliconised polyester[1,6] PVC or acrylic. PVC and PVF2 are the most common in UK sheet roofing, and are reported to be the most durable.[6] While these organic coatings extend the lifespan of the steel, they have serious environmental impacts during manufacture, particularly PVC, the manufacture of which results in the release of dioxins and plasticisers, which are reported to act as hormone disrupters.

Embodied Energy

For ease of comparison, all embodied energy figures in this issue have been converted to MJ kg⁻¹ (megajoules per kilogram).

In order to give a realistic idea of the embodied energy of roofing products, we have attempted to calculate the energy required per unit area of roofing. (see table below)

This will reflect the fact that while some products such as aluminium may have a high embodied energy per kg, a lower mass may be required than for a lower embodied energy product such as clay tiles.

The sources of the embodied energy figures in MJ/kg are listed in the relevant sections of the Product Analysis.

Weights per m² were kindly supplied by Carl Thompson Associates, Birkenhead, or obtained directly from manufacturers information packs.(refs 72, 73, 74)

We have given a maximum and minimum figure where applicable to reflect the range of weights and embodied energy of products.

The energy multiplier has been included for ease of comparison, and is derived using the same method as for the Unit Price Multiplier.

Material	Energy MJ/kg	Weight as Roofing kg/m2	Energy MJ/m2	Energy Multiplier (Slate = 1)
Clay Tiles	6.3	42 - 68	265 - 428	2.1 - 3.3
Natural Slate	<4	<32 - 40	<128 - 160	1 - 1.3
Concrete Tile	1	43 - 87	43 - 87	0.3 - 0.7
Fibre-Cement Tile	-We were unable to find reliable embodied energy figures for these products.	19 - 22	Likely to be lower than concrete tile, due to lower mass required per m2	-
Reconstructed Slate		17 - 20		-
Polymer Modified Cement		-	-	-
Asphalt Shingles (Organic Mat)	-	-	283	2.2
Asphalt Shingles (Fibreglass Mat)	-	-	>283	>2.2
Steel (0.7mm)	25 - 33	7.3 - 8.7	181 - 287	1.4 - 2.2
Stainless Steel	11	3	33	0.3
Aluminium (0.9mm)	180 - 240	3 - 5.1	544 - 922	4.2 - 7.2
Aluminium (Recycled) (0.9mm)	10 - 18	"	30 - 92	0.2 - 0.7
Lead	190	15 - 40	2848 - 7663	22.3 - 59.9
Lead (Recycled)	10	"	150 - 403	1.2 - 3.1
Copper	70	2.8 - 6.3	197 - 441	1.5 - 3.4

12.3 Best Buys

Best Buy:	Reclaimed Tiles/Slates Certified Wooden Shingles (see p.157)
Second Choice:	Natural Slates
Third Choice:	Clay/Cement based tiles

12.3.1 A Note on Methodology

As in previous chapters, we have attempted to review the impacts of some multi-component products by combining the impacts of each component, the value of each component impact roughly weighted depending on its proportion of the total product. For example, in the product table, the global warming potential of resin/polymer bonded slate is derived from the separate impacts of the cement, glass fibre, polymer and slate aggregate in the product. Rather than simply adding the impacts together, an attempt has been made to recognise that each component only makes up a certain proportion of the whole. For example, the inclusion of slate aggregate in polymer bonded cement reduces the requirement for the other, higher impact components, such as cement and glass fibre, reducing the overall impact.

The exception to this system is the impact of organic coatings for sheet steel roofing, which have been assessed separately.

The environmental best buy is locally reclaimed durable roofing tiles - natural slate, clay, concrete, fibre cement, polymer bonded and polymer modified cement. Although these all have impacts during manufacture, using reclaimed materials avoids the impacts of manufacturing the new product, and reduces the waste burden.

(a) Tiles

The environmental best option in the tile section is probably locally produced slate from 'environmentally sympathetic' quarries (see Suppliers section, p.167) due to its relatively low manufacturing impacts, although embodied energy and resource use are higher than would be expected, due to high wastage rates. Slate is also a fairly high cost option. Clay and cement based tiles are probably the next best option. Of the two groups, clay has the highest embodied energy, but comes out more favourably in terms of its other impacts. The more manufactured cement based tiles (fibre-cement, reconstructed slate and polymer bonded slates) all contain high energy, high impact materials such as glass fibre and organic polymers, although the impacts of these are offset somewhat by a reduction in cement use and lower tile weight. Resin or polymer bonded slate slabs have an apparent environmental advantage in that they utilise poorer quality crushed slate and stone waste from the manufacture of finished slate products. On the one hand, this resource saving over natural slate is at the expense of the use of manufactured fibres, petroleum based polymers and cement, which are made from non-renewable resources and have additional impacts over those of finished slate tiles. On the other hand, in comparison with products made entirely with 'manufactured ' materials, resin or polymer bonded slates could be seen as environmentally advantageous as most of their mass is composed of 'waste' slate, with the more environmentally damaging manufactured materials comprising only a small proportion of the total product.

(b) Sheet Metal

Of the sheet steel products, stainless steel appears to be a good choice, despite its higher cost. From a resource use view, stainless steel may be an excellent choice as it is claimed to be manufactured using recycled steel[17] and is itself readily recycled. It is also more durable than mild steel and has a much lower embodied energy, although there are greater problems of toxic metal release during production due to the use of heavy metals in the alloy.[17]

All of the products reviewed require the use of non-renewable resources, and with the exception of slate, all have fairly high manufacturing impacts.

With this in mind, the most important aspects of a roofing product in terms of environmental impact may be its durability and reusability (classed under the 'recycling' heading).

With the exception of the sheet metal products, none of the products reviewed are recyclable except as aggregate, fill or low grade products (eg, asphalt tiles can be used to patch roads). However, most are reclaimable - the potential for reclamation determined mainly by their durability (a smaller dot indicates higher potential for reclamation).

Another important factor is materials efficiency. Well designed interlocking slates and tiles which have a small overlap between units will use less material, and are therefore environmentally preferable to tiles which require a large overlap between units.

A range of 'Environmentally Sound' materials can be found in the Alternatives section, page 164.

Energy use ratings are for energy per m² for roofing, rather than energy per unit weight. Detail of how figures were obtained are given in table 1, p.146

	£	Production								Use				
	Unit Price Multiplier	Energy Use	Resource Depletion (non-bio)	Resource Depletion (bio)	Global Warming	Ozone Depletion	Toxics	Acid Rain	Occupational Health	Recycling/Reuse/Disposal	Durability/Maintenance	Health	Other	ALERT!
'Natural' Tiles														
Clay Tile	0.8-2	●	·		●		●	●					·	
Natural Slate	2-4.6	·	●	·	·								·	
Cement Based Tile														
Concrete Tile	0.6	·	●	·	●		●	●	●	●		·	●	Haz. Waste?
Fibre Cement Tile — Glass Fibre	1	·	●		●		●	●	●	●		·	●	Haz. Waste?
Fibre Cement Tile — Synthetic Fibre	1	·	●		●		●	●	●	●		·	●	Haz. Waste?
Fibre Cement Tile — Cellulose Fibre	1	·	●	·	●		·	·	●	●		●	●	
Resin Bonded (reconstructed) Slate	1-1.6	·	●	·	●		●	●	●	●		·	●	Haz. Waste?
Polymer Modified Cement Slates	-	·	●		●		●	●					●	Haz. Waste?
Ferrocement	-	·	●		●		●	●	●	●		·	●	Haz. Waste?
Asphalt Shingles														
Organic/Cellulose Mat	-	●	●		·		●	●	·	●		●	●	
Glass Fibre Mat	0.6	●	●		●		●	●	·	●		●	●	
Metal Sheet														
Steel Sheet — Alu. Coated	-	●	·	●	●		●	·		●		·		H.D.
Steel Sheet — Galvanised	0.7	●	●	·	●		●	●		●		●		H.D.
Additional Impacts of Organic Coatings for Steel Sheet — PVC	-	·	●		●		●	●		●		·		H.D.
Additional Impacts of Organic Coatings for Steel Sheet — Polyester	-	●	●		●		●	●		·		●		
Additional Impacts of Organic Coatings for Steel Sheet — Acrylic	-	●	●		●		·	●	●	·		●		
Stainless Steel Sheet	2.4	·	·		●		●	●				·		H.D.
Aluminium Sheet	1.4	●	·	●	·		●	·				·		
Lead Sheet	2-3.7	●	●	·	●		●	●	·		●			

H.D. = Hormone Disruptor

Unit Price Multiplier

The unit price multiplier in this issue is for materials plus labour cost per m². This is in order to account for materials which have a low purchase cost, but high labour costs, such as copper sheet. The multiplier is derived mainly from prices listed in Spons Architects and Builders Price Book,[61] and calculated using the cost of 'Eternit' fibre-cement tiles (21.58m² inc. labour) as a base cost.

12.4 Product Analysis

12.4.1 'Natural' Tiles

(a) Clay Tiles

Manufacture

Energy Use

Based on data from other vitreous clay products, unglazed clay tiles require about 6.3MJ/kg - an estimated 250MJ/m^2 for roofing.[1] Due to the high weight of clay, transportation will also consume significant amounts of energy unless local clay sources and manufacturers are used. The quality of the tile depends on the temperature it is fired at, as well as the quality of the clay.[1] There is therefore a trade off between durability and the larger amount of energy required to produce high quality tiles.

Resource Use (non-bio)

Clay is an abundant resource in most of the world.[1]
Barium carbonate is often added to prevent effluorescence.[1]

Resource Use (bio)

Clay pits temporarily affect ecology but are usually restored[37] and so this is only of importance if ecologically rare or sensitive areas are excavated. Clay pits are often 'restored' to wetlands, which could be seen as an ecological advantage.

Global Warming

NOx, a greenhouse gas, is released during firing.[37]

Acid Rain

SO_2 and NOx, which contribute to acid rain, are released during firing of clay products.[37]

Toxics

Emissions to air during production include fluorine and chlorine compounds.[37] Emissions to air and water during extraction are mainly inert particulates.[37] Material from flue gas treatment systems, disposed to land, may contain halogens (chlorine, bromine, fluorine etc).[37] Emissions from clay firing are controlled under the Environmental Protection (Prescribed Processes and Substances) Regulations 1991.[38]

Other

Excavation of clay can cause a local nuisance.[(37)]

Use

Durability

Clay tiles are amongst the most durable of building materials.[33] The more expensive products tend to be more resistant to repeated freeze-thaw cycles and can last 100-125 years of repeated freeze-thawing. Tiles with higher water absorption are more susceptible to frost damage.[(1)] (See Energy Use above)

Recycling/Reuse/Disposal

Due to their high durability, clay tiles are extensively reclaimed for reuse,[32] and use of reclaimed clay tiles is recommended as an environmental option.

(b) Slates

Manufacture

Energy Use

Energy use is limited to the fossil fuels used in quarrying, shaping and transporting the slate.[1] Slate from a local source will therefore have the lowest energy cost. Because of high wastage rates, the energy requirements for finished stone products can be up to 4 MJ/kg of stone produced,[37] but may be considerably less depending on extraction methods.

Resource Use (non-bio)

We were unable to find data for remaining slate reserves. Because of the requirement to produce stone of a given dimension and quality, reject material can form a large percentage of quarry production.[37]

Resource Use (bio)

Much of the slate in the UK is located in areas of great natural beauty, such as the Lake District and North Wales,[33] and quarrying has the potential to cause significant landscape and ecological damage, unless carried out with extreme care.

Global Warming

It is estimated that CO_2 emissions from production plant, quarry transport and electricity are of the order of 0.53 tonnes CO_2 per tonne of finished product.[37] This figure does include transport to the point of use.

Other

Slate quarrying can cause local impacts such as noise, dust and vibration, plus increased heavy road traffic.[37]

Use

Durability

Slates are highly durable, and good slate such as that complying with BS 680 Part 2, is one of the most durable building materials. Poor slate, however, may begin to decay within a few months, especially in damp conditions in industrial areas.[31]

Recycling/Biodegradability

Their extremely high durability allows active reclamation and reuse of good quality slates, which may outlast the building structure.[37] Roofing from houses over 100 years old has been successfully removed and reused, and the market price for recycled slate is only slightly lower than for new.[1] The use of reclaimed slates is recommended as an environmental option.

Researchers at Argonne National Laboratory, USA, have found a way to make concrete reabsorb nearly 80% of the CO_2 released, by curing in a CO_2 rich environment. This also speeds up the concrete curing time, and improves strength. The potential for this outside the laboratory is not yet known.[30]

Other

Originally a locally based industry, with slates obtained from local quarries, the slate industry is now global with slates being imported from South America, Asia and Africa, taking advantage of cheap labour costs.[58] This will add to the energy use due to long distance transportation - although this may be offset by more labour intensive quarrying methods - and may lead to greater local environmental degradation due to the less strict environmental legislation in some developing countries. Slate quarrying can cause local impacts such as noise, dust and vibration, habitat destruction, visual pollution and increased heavy road traffic.[37]

12.4.2 Cement Based Tiles

(a) Concrete Tiles

Manufacture

Energy Use

Ordinary Portland Cement (OPC), which accounts for 25% of the weight or concrete tiles, is an energy intensive material[1] with an embodied energy of 6.1MJ kg^{-1} (wet kiln production) or 3.4MJ kg^{-1} (dry kiln production).[37] The embodied energy of concrete is approximately 1MJ kg^{-1}, lower than that of OPC due to its high sand content.[37] Energy is also consumed in tile manufacture, which involves high pressure extrusion, plus 'curing' at 40°C in high humidity for 8-24 hours.[33] The polymer emulsions and acrylic paints applied to suppress effluorescence, provide colour and help seal sanded and textured surface finishes, are manufactured from petrochemicals, which have high embodied energy.[13]

Resource Use (non-bio)

The raw materials for cement manufacture are limestone/chalk and clay/shale,[37] which are abundant in the UK, although permitted reserves are running low in some areas, notably the south-east.[63] The use of pulverised fuel ash and granulated blast furnace ash in concrete, by-products of the power generation and iron/steel industries respectively, has increased significantly over the last 10-12 years,[37] reducing the amount of quarried material required.

Global Warming

The manufacture of Portland cement releases around 500kg CO_2 per tonne,[30] and is the only significant producer of CO_2 other than fossil fuel burning, responsible for 8-10% of total emissions.[16] Some of this is re-absorbed during setting.[30]

Toxics

OPC contains heavy metals, "of which a high proportion are lost to atmosphere" on firing.[64] Organic hydrocarbons and carbon monoxide are also released, and fluorine can also be present.[42]

Acrylonitrile, used in the manufacture of acrylic paints used to prevent effluoresence, is a suspected carcinogen

and is reported to cause headaches, breathing difficulties and nausea.[9,10]

Acid Rain

Burning fuels to heat cement kilns releases NOx and SO_2. NOx is released to atmosphere, whereas most of the SO_2 is reabsorbed into the cement.[42]

Other

Cement manufacture results in the production of significant amounts of dust, which can be hard to control.[37] Extraction of the raw materials for OPC and sand can also cause localised problems of noise, vibration and visual impact.[34] Admixtures, added in small quantities to concrete or mortar in order to alter its workability, setting rate, strength and durability,[65] are products of the chemical industry, or by-products of wood pulp manufacture.[37] These are listed below;

Cement Admixtures

chloride
ethyl vinyl acetate
formate
hydroxycarbolic acid
lignosulphonate
melamine condensate
naphthalene condensate
phosphates
polyhydroxyl compound
polyvinyl acetate
stearate or derivative
styrene acrylic
styrene butadiene
wood resin derivative.[40]

The only published study to date on the environmental impacts of these is 'Concrete Admixtures and the Environment' (Industrieverband Bauchemie und Holzschutzmittel, Frankfurt, 1993), for which an English translation is planned. The European Federation of Concrete Admixtures Associations is currently carrying out a study on the impacts of admixtures, but no conclusions have been published so far.

Use

Durability

Concrete is durable and resistant to surface temperatures between -20°C and 70°C,[33] with a lifetime in excess of 60 years.[37] However, it is suggested that concrete tile is less durable than the clay it displaces from the roofing sector.[1] Problems of surface corrosion[33] and discoloration[37] may be accelerated in areas of high industrial pollution where SO_2 levels exceed 70ugm^{-3} of air.[33]

Recycling/Reuse/Disposal

Of the 0.36 million tonnes of concrete roof tile demolition waste, 0.2 million tonnes are recycled.[41] However, most of this is recycled as fill and sub-base material, rather than reclaimed for re-use in roofing.[37] Reclaimed concrete tiles are recommended as an 'environmental' option.

ALERT

See 'Waste Incineration in Cement Kilns' , chapter 6.

(b) Fibre-Cement Tiles.

Impacts relating to the use of Ordinary Portland Cement are detailed in the concrete tile section above. Only the impacts which differ from those of concrete tiles are detailed in this section. The impacts of fibre manufacture are dealt with separately.

Manufacture

Energy Use

Use a smaller amount of energy intensive cement than concrete tiles.[1]

Some manufacturers cure the tiles in an autoclave to speed the process, which increases the embodied energy over that of the air dried product.

Resource Use (non-bio)

Fibre-cement tiles use less cement than concrete tiles.[1]

Toxics

Until the discovery of its carcinogenic properties, all synthetic slates were produced using cement bonded with asbestos fibres.[33] Asbestos has since been replaced by synthetic fibres or natural fibres such as sisal, and filling compounds.[33] The impacts of these are discussed below. Acrylonitrile, used in the manufacture of acrylic paints used to prevent effluoresence, is a suspected carcinogen and is reported to cause headaches, breathing difficulties and nausea.[9,10]

Use

Durability

Fibre-cement tiles are rot proof and resistant to insect and vermin attack; Products tested to BS4624 are unaffected by frost.

Some manufacturers cure fibre-cement tiles using an autoclave in order to reduce the curing time. This makes the tiles more brittle than air-cured tiles,[1] as well as increasing the energy use during manufacture.

Tiles may lighten in colour due to UV radiation, and in heavily polluted areas, slight surface softening of the cement may occur.[33]

Cellulosic fibres used to substitute asbestos, have increased susceptibility to freeze-thaw damage.[1]

Recycling/Reuse/Disposal

Fibre-cement tiles can be reclaimed for re-use due to their high durability, unless they contain asbestos, which would make them unsafe for reuse.

ALERT

See Cement Kilns, chapter 6

THERE NOW FOLLOWS A CONCISE AND DEFINITIVE EXPLANATION AS TO WHY WE NEED TO IMPORT NEWLY MANUFACTURED SLATE TILES INTO THIS COUNTRY FROM ANOTHER ONE THAT IS THOUSANDS AND THOUSANDS OF MILES AWAY INSTEAD OF RECLAIMING PERFECTLY GOOD ONES FROM LOCAL BUILDINGS THAT ARE ABOUT TO BE DEMOLISHED...

(c) Additional Impacts of Fibres in Fibre-Cement

Glass Fibre

Manufacture

Energy Use

The embodied energy of glass fibre is estimated at 15 - 18MJ kg^{-1}.[1, 37] This is higher than that of cement, and therefore increases the overall embodied energy of the product.

Resource Use (non-bio)

The raw materials for fibreglass manufacture are silica sand, boron, limestone[1] and sodium carbonate[37], which are non-renewable resources.[46]

Sodium carbonate for glass manufacture can be mined, or produced by the 'Solvay process', which involves the addition of ammonia (produced from natural gas, a limited resource) to brine (produced from mined rock salt).[37]

Global Warming

Gaseous emissions from fibreglass production include the 'greenhouse' gases NOx, CO_2 and carbon monoxide.[37]

Acid Rain

Sulphur and nitrogen oxides which form acid rain are produced during fibreglass manufacture.[37]

Toxics

Emissions to air from fibreglass manufacture include fluorides, chlorides and particulates (including glass fibres). Solid wastes include organic solvents, alkalis and 'alkali earth' metals.[37]

Acid Rain

Glass fibre manufacture contributes to acid rain formation, mainly through the fuels used to melt the ingredients.[66]

Occupational Health

Dust arising from glass fibre processes can cause skin, throat and chest complaints[66] and in the USA, glass fibres narrowly escaped being listed by the government as carcinogenic due to corporate lobbying.[67] However, it appears that the main health risks are associated with insulation fibres, which come in much smaller sizes than structural glass fibres.[62]

Other

Quarrying sand for glass manufacture can have localised impacts of noise, vibration, visual and dust pollution, and habitat destruction.[37] Extraction of natural gas, used in the Solvay process, can also have serious environmental impacts.[25]

Use

Durability

see fibre cement tiles section above

Recycling & Biodegradability

see fibre-cement tiles section above

Synthetic Polymer Fibre

Polyvinyl alcohol (PVAL) is the most commonly used in fibre reinforced cement roofing, and polyacrylonitrile (PAN) is sometimes used.[47] Polypropylene fibres are used to make strong extruded slate.[47]

Manufacture

Energy Use

Plastic polymers are produced using high energy processes, with oil or gas as raw materials, which themselves have a high embodied energy.[13] Polypropylene has an embodied energy of 100MJ kg^{-1}.[12]

Resource Use (non-bio)

Oil is the main raw material used in the manufacture of synthetic fibres. This is a non-renewable resource.

Global Warming

Petroleum refining and synthetic polymer manufacture are major sources of NOx, CO_2 methane and other 'greenhouse' gases.[16, 25]

Acid Rain

Petrochemicals refining and synthetic polymer manufacture are major sources of SO_2 and NOx, the gases responsible for acid deposition.[13, 16]

Toxics/Occupational Health

The petrochemicals industry is responsible for over half of all emissions of toxics to the environment.[16]

Acrylonitrile, the monomer used in the production of Polyacrylonitrile, is a suspected carcinogen and is reported to cause headaches, breathing difficulties and nausea.[9,10]

Asbestos

Asbestos is mined from asbestos containing rock, which is crushed and milled to produce raw asbestos. The principal suppliers were Canada, the former Soviet Union and Southern Africa.[47] Its use is still permitted in this country, provided the fibres are bound within the matrix of the product[46] However, it is discouraged on health grounds, and several alternatives are available.

Asbestos can be a major health hazard during mining, manufacture, demolition and refurbishment.[46] The inhalation of fibres can cause asbestosis (scarring of lung tissue), lung cancer and a number of other diseases of the lungs and chest.[46] When used in roof sheeting/tile, asbestos presents a lower risk than loose fibres, because the cement binding prevents the loss of fibres. However, loss may occur when the material is cut, drilled or broken in any way.[46] There is no evidence that ingestion, as opposed to inhalation, represents a health hazard.[46]

Use

Health

We found no evidence of a health risk during use from any of the synthetic fibres reviewed in this report.

Durability

Synthetic fibre products are less strong than asbestos fibre cement tiles.[47] Long term durability is difficult to assess in new products, but despite problems with alternatives developed in the early eighties, products developed within the last 8 years seem to be performing satisfactorily.[47]

Recycling/Reuse/Disposal

Synthetic polymers are highly persistent in the environment[49] making their disposal a problem. Synthetic fibre cement tiles are durable, and therefore can be reclaimed and reused.

Other

The extraction, transport and refining of oil required for the production of synthetic fibres can have enormous localised environmental impacts.[13]

Cellulose Fibres

Manufacture

Energy Use

We found no data regarding energy use for cellulose fibres. However, as these are generally natural fibres (wood fibre, sisal, coir etc.)[36] rather than manufactured products, their embodied energy is likely to be significantly lower than glass or synthetic fibres.

Resource Use (bio)

Cellulose fibres are generally derived from renewable biological resources, the most common fibres being sisal and coir.[36]

Durability

Sisal and coir fibre reinforced cement tiles have a shorter life expectancy than those made
with 'industrial' fibres, and can be expected to last between 10 and 15 years.[36] However, many fibre-cement tile manufacturers extend their lifespan considerably, using a blend of natural and manufactured fibres.[60]

(d) Resin and Polymer Bonded Slates

Manufactured using natural aggregates of crushed slate and stone, reinforced with glass fibre in a polyester or acrylic resin binder.[33]

Manufacture

Impacts relating to ordinary glass fibres are detailed in the fibre cement tile section on page 151.

Energy Use

The material is moulded at high temperature and pressure, to give the appearance of slate.[33]
The embodied energy of glass fibre is estimated at 15 - 18kJ kg^{-1}.[1, 37]

Synthetic polymers are produced using high energy processes, with oil or gas as raw materials, which themselves have a high embodied energy.[13]
Energy use for the slate aggregate is limited to the fossil fuels used in quarrying, crushing and transporting the slate.[1]
Also see 'Concrete tiles' section (p.150) for the impacts of Portland Cement.

Resource Use (non-bio)

Oil is the main raw material for acrylic and polyester resins. This is a non-renewable resource.
The crushed slate/stone used as an aggregate is likely to be waste material from the quarrying and manufacture of finished slate products.
Also see 'Glass Fibre' section opposite.

Resource Use (bio)

Much of the slate used as aggregate is located in areas of great natural beauty, such as the Lake District and North Wales,[33] and quarrying has the potential to cause significant landscape and ecological damage, unless carried out with extreme care.

Global Warming

Synthetic polymer manufacture is a major source of 'greenhouse' gases.[16,25]
Also see 'Glass Fibre' section opposite.
It is estimated that CO_2 emissions from producion plant, quarry transport and electricity during slate production are of the order of 0.53 tonnes CO_2 per tonne of finished product,[37] although this is likely to be less for crushed aggregate where there is less wastage of material. This figure does include transport to the point of use.

Acid Rain

Synthetic polymer manufacture is also a major source of SO_2 and NOx.[13,16]
Also see 'Glass fibre' section opposite

Toxics

The petrochemicals industry, from which the synthetic binders are derived, is responsible for over half of all emissions of toxics to the environment.[16]
Acrylonitrile, used in the manufacture of acrylic resins and acrylic paints used to prevent effluoresence, is a suspected carcinogen and is reported to cause headaches, breathing difficulties and nausea.[9,10] Although associated with toxic emissions, polyester has a relatively small impact when compared with other plastics such as PVC.[11]
Also see 'Glass Fibres' section opposite

Use

Occupational Health

Acrylonitrile, used in the manufacture of acrylic resins and acrylic paints used to prevent effluoresence, is a suspected carcinogen and is reported to cause headaches, breathing difficulties and nausea.[9,10]

Health

We found no evidence of a health risk during use from any of the synthetic binders reviewed in this report.

Durability
Resin bonded slates have a high durability, and are highly resistant to airborne pollution.[33]

Other
Fire resistance is only to class-2 fire rating, making these tiles unsuitable for certain locations.[33] The extraction, transport and refining of oil, used to produce the synthetic bonding material, can have enormous localised environmental impacts.[13]

Slate quarrying for the aggregate portion of the tile, and sand quarrying for glass fibre manufacture, can cause local impacts such as noise, dust and vibration, habitat destruction, visual pollution and increased heavy road traffic.[37]

Extraction of natural gas, used in the Solvay process during glass fibre manufacture, can also have serious environmental impacts.[25]

ALERT
See 'Cement Kilns - The Burning Issue' box, chapter 6

(e) Polymer Modified Cement Slates

An adaptation of cement slates, whereby part of the mixing water is replaced with an aqueous polymer emulsion. This improves the workability of the cement, improves flexured load carrying capacity and improves chemical and frost resistance. This allows the manufacture of extremely strong lightweight tiles, with lower surface erosion than standard concrete.

Impacts are as for Portland Cement (see concrete tiles section, p.150), with the following modifications for the 5% polymer content;

Energy Use
Synthetic polymers are produced using high energy processes, with oil or gas as raw materials, which themselves have a high embodied energy.[13]
(Also see 'Concrete tiles' section p.150)

Resource Use (non-bio)
Oil is the main raw material for most synthetic polymers. This is a non-renewable resource. The polymer content is around 5%, and acrylic paints are often used to provide colour, additional protection and prevent effluoresence.[33]
(Also see 'Concrete tiles' section, p.150)

Global Warming
Synthetic polymer manufacture is also a major source of 'greenhouse' gases.[16,25]

Acid Rain
Synthetic polymer manufacture is a major source of SO_2 and NOx.[13,16]

Toxics
The petrochemicals industry, from which the synthetic binders are derived, is responsible for over half of all emissions of toxics to the environment.[16]

Acrylonitrile, used in the manufacture of acrylic paints used to prevent effluoresence, is a suspected carcinogen and is reported to cause headaches, breathing difficulties and nausea.[9,10]

Use
Durability
The polymer improves the flexured load carrying capacity and improves chemical and frost resistance. This allows the manufacture of extremely strong lightweight tiles, with lower surface erosion than standard concrete.[33]

Other
Unlike resin- and polymer-bonded slates, polymer modified slates can achieve class-0 rating for spread of flame.[33]

(f) Ferro-Cement Roofing Element

Ferro-cement is a form of reinforced mortar, in which closely spaced and evenly distributed thin wire meshes are filled with rich cement-sand mortar.[55]

The impacts of cement are discussed in the Concrete section above; the impacts of the steel mesh are discussed in the steel sheet roofing section, p.156

12.4.3 Asphalt Tiles

(a) Asphalt Tiles (Organic Matting)

Manufacture

Energy Use

Energy use in asphalt shingle manufacture is fairly high - the embodied energy for organic shingles has been estimated at 284,000kJ m^{-2}.[1]

Resource Use (non-bio).

Asphalt is made from low grade products of petrol refining (bitumen), and is a limited resource.[1, 37] Manufacturing organic shingles requires significantly more asphalt than fibreglass shingles.[1]

Rock dust used as a mineral filler, and mineral granules used as a surface coating are also non-renewable resources, obtained by quarrying - although these are not in limited supply. Some companies (eg CertainTeed) use granules made from coal-fired boiler slag, an industrial waste.[1]

Resource Use (bio)

Organic felts use recycled materials, often including recycled paper fibres.[1] This saves on both materials and energy when compared to the use of virgin resources.[39]

Global Warming

Oil extraction and petrochemical refining are major sources of CO_2, NOx, methane and other 'Greenhouse' gases.[16,25,37]

Acid Rain

Oil extraction and petrochemical refining are major sources of SO^2 and NOx, which form acid rain.[16,25,37]

Toxics

The petrochemicals industry is responsible for over half of all emissions of toxics to the environment.[16] Solid wastes from refining and extraction include polynuclear aromatics and heavy metals.[37]

Occupational Health

The main impact of asphalt tiles during construction is odour from volatile organic compounds.[37]

Use

Durability

Asphalt tends to degrade on exposure to sunlight[37] although coatings can reduce this effect.[1] Asphalt tiles are often considered as a 'throwaway' roofing, due to their low durability. Budget (20 - 25 year) shingles are not recommended as a 'good environmental choice' because of the environmental cost of frequent replacement.[1] Organic mats have a higher tear strength than fibreglass mats.[1]

Recycling/Reuse/Disposal

In the USA, ReClaim uses shingles to make pavement patching materials and paving material for low-traffic areas.[1] Due to their low durability, the potential for reclamation and reuse of asphalt tiles as roofing elements is likely to be low.

Other

The extraction, transport and refining of petroleum products can have enormous localised impacts in the event of accidental and operational spills.[13,37]

(b) Asphalt Tiles (Fibreglass Matting)

To save repetition, this section only covers areas in which fibreglass mat asphalt tiles have an impact which is different to organic mat asphalt tiles. Otherwise, impacts are the same as detailed in the previous analysis.

Energy Use

Energy use for fibreglass backed shingles is likely to be higher than for natural fibre mats, due to the high embodied energy of fibreglass, estimated at 15 - 18MJ kg^{-1}.[1, 37]

Resource Use (non-bio)

The raw materials for fibreglass manufacture are silica sand, boron, limestone[1] and sodium carbonate,[37] which are non-renewable resources.[46] Little if any recycled glass cullet is used in the manufacture of fibreglass for asphalt shingles.[1]

Sodium carbonate for glass manufacture can be mined, or produced by the Solvay process, which involves the addition of ammonia (produced from natural gas, a limited resource) to brine (produced from mined rock salt).[37]

Global Warming

Gaseous emissions from fibreglass production include the 'greenhouse' gases NOx, CO_2 and carbon monoxide.[37]

Toxics

Emissions to air from fibreglass manufacture include fluorides, chlorides and particulates (including glass fibres). Solid wastes include organic solvents, alkalis and 'alkali earth' metals.[37]

Acid Rain

Sulphur and nitrogen oxides which form acid rain are produced during fibreglass manufacture.[37]

Use

Occupational Health

See 'Glass Fibre' section, p.152.

Recycling & Biodegradability

See 'Asphalt Tiles - Organic Matting' section above.

Durability

Budget (20 - 25 year) shingles are not recommended as a ' good environmental choice' because of the environmental cost of frequent replacement, although more expensive shingles have higher durability.[1] Fibreglass mats have a lower tear strength than organic mats.[1]

Other

Quarrying sand for glass manufacture can have localised impacts of noise, vibration, visual and dust pollution, and habitat destruction.[37] Winning of natural gas, used in the Solvay process, can also have serious environmental impacts.[25]

12.4.4 Sheet Metal Roofing

(a) Steel Sheet

Manufacture

Iron and steel production has been ranked as the second most polluting industry in the UK, second only to coke production (of which it is a major customer).[83]

Energy Use

The embodied energy of steel is 25-33MJ kg^{-1}.[(17)] (19,200 Btu/lb.)[5] 99.7% of principle feedstocks used in our steel industry is imported, mainly from Australia, the Americas and South Africa,[19] and the transport energy costs should be taken into account.

Resource use (non-bio)

Iron-ore, limestone and coke (made from coal) are required for iron and steel production. Easily available sources of iron ore are getting scarce, and the raw materials have to be transported increasing distances.[84]

Proven reserves of steel are estimated to be sufficient for 100 years supply if demand continues to rise exponentially, and 200 years at current levels of demand.[20] Steel is manufactured using about 20% recycled content, 14% of which is post-consumer[1].

Resource Use (bio)

Brazil exports huge quantities of iron to the West, much of it produced with charcoal made from rainforest timber.[34,64] Clearance of land for iron ore extraction in Brazil may also contribute to rainforest destruction.[17]

Global Warming/Acid Rain

Combustion emissions from ore refinement, blast furnace and oxygen furnace operations include greenhouse- and acid rain forming gases. About 3 tonnes of CO_2 are emitted per tonne of steel produced from ore, and 1.6 tonnes per tonne of recycled steel.[17] Due to the size of the industry, global figures for CO_2 emissions from iron & steel production are significant, although much smaller than those from burning fossil fuels (about 1.5%). CO_2 emissions incurred during global transport of raw materials (See 'energy use') should also be considered.[17]

Acid Rain

Emissions of sulphur dioxide and nitrogen oxides are associated with iron and steel production.[16,78]

Toxics

Estimates from the Department of the Environment rate sinistering (an early stage of iron and steel production) as possibly one of the largest sources of dioxin emissions in the UK, but there are no reliable figures as yet.[68] Steel smelting is also listed as a major source of dioxin, as a result of the recycling of scrap steel with PVC and other plastic coatings.[69, 70]

Volumes of dust are produced by ore refinement and blast furnace operations to produce raw iron.[5] There is also a danger of water pollution from improper disposal of processing waters from mining and milling operations.[5] Iron ores are relatively innocuous, but toxic metals are

released in low concentrations as solid and liquid waste during refining.[17] In the UK, emissions to air are controlled within HMIP limits, although some pollutants may still be released, including heavy metals, coal dust, oils, fluorides, carbonyls, fluorides, alkali fume, dust and resin fume.[17] Metal smelting industries are second only to the chemicals industry in terms of total emissions of toxics to the environment.[16] In the UK, iron and steel production tops the chart of fines for water pollution offences.[85]

Other

The extraction of iron ore, limestone and coal all has an impact locally.[84]

Use

Durability

Steel roofing can be expected to last around 25-30 years.[2] A service life of over 50 years is unusual, except for the most expensive products.[1] Durability is dependent mainly on the galvanising or organic coating material (see below).

Recycling/Reuse/Disposal

The ease of reclamation is the main environmental advantage of steel roofing.[1] Steel is easily removed from the waste stream magnetically and can be recycled into high quality steel products,[5] and the estimated recovery rate is currently 60-70%.[17] Recycled steel consumes around 30% of the energy of primary production (8-10MJ kg^{-1} recycled), including the energy required to gather the scrap for recycling.[17] It is thought that through increasing the extent of recycling and using renewable energy, there is scope for steel to be produced sustainably.[17]

ALERT

Dioxins, released during the manufacture and recycling of steel, have been identified as hormone disrupters.[57] -See 'Alert' p.174 for further detail.

Aluminium Coating

See 'Aluminium Sheet' section below, (bearing in mind that the amount of aluminium used in a protective layer over steel will be considerably less than that used in aluminium sheeting).

Durability

Aluminium coatings provide greater protection than zinc as they are less easily scratched.[1] While zinc coatings are generally 'sacrificial' (ie, they corrode preferentially to the underlying steel), aluminium coatings form an impermeable oxide surface sheath, which prevents further corrosion.

Zinc Galvanising layer

Galvanising is applied to steel sheet in manufacture to thicknesses of about 3 to 10um; alternatively, finished products may be hot dip galvanised by lowering into a bath of molten metal, resulting in a thickness of 20 to 50um.[17]

POLITICAL ROOFING TRENDS 1979 ᴛᴏ 1996

THATCHED

SCARGILLED

ASHDOWNED

BLAIRED

Manufacture

Energy Use

Most of the zinc used in the UK is imported from Australia, Peru and the USA,[18] which has implications for transport energy requirements. Processing is energy intensive,[46] with total energy in primary production estimated at 65MJ kg[-1], 86% in smelting and refining.[17]

Resource Use (non-bio)

Zinc is a non-renewable resource.[46] We found no figures to indicate the size of remaining exploitable reserves.

Resource Use (bio)

Quarrying for metal ores can result in the destruction of wildlife resources and habitat loss.[17]

Global Warming

CO_2 emissions are estimated at 6 tonnes per tonne of zinc produced.[17]

Acid Rain

SO_2 and NOx emissions will be "substantial" owing to the fossil fuels consumed during zinc manufacture.[17]

Toxics

'Passivisation' of the underlying steel to prevent 'white rusting' involves dipping in a chemical solution, frequently based on chromates. These solutions produce highly toxic waste products.[46]

Extracting and enriching zinc ore releases toxic and phytoxic (toxic to plants) lead, antimony, arsenic and bismuth.[17,21] These may bioaccumulate in crops, particularly root crops, giving rise to a potential health hazard.[17] Disposal of waste waters from the enrichment process and drainage water run-off to rivers and groundwaters from stored ores is a potential source of water pollution, although these can be controlled by biological effluent treatment.[17] Acid mine drainage, containing dissolved metals, can have serious impacts on aquatic flora and fauna.[21] Tailings, contaminated with surfactants or acids and heavy metals must also be disposed of.[17] This is generally carried out using tailings lagoons, which can take up a large land area and are difficult to revegetate due to instability and phytoxicity. In the UK, emissions to air are controlled within HMIP limits under the 1990 Environmental Protection Act,[23] although some pollutants may still be released, including heavy metals, coal dust, oils, fluorides, carbonyls, fluorides, alkali fume, dust and resin fume.[17]

Durability

Zinc galvanising provides corrosion protection to steel roofing sheet and tile, thus extending its expected life.[46] However, zinc coatings are prone to chipping and erosion, allowing corrosion of the underlying steel.[36]

Recycling/Reuse/Disposal

Zinc is extensively recycled in alloys with copper and other metals,[17] although zinc from galvanising coatings is not readily recycled. In the US, zinc coatings on sheet steel are usually removed and recycled although aluminium coatings are not.[1]

Recycled zinc has a much lower impact than production from the ore.[17]

Organic Coatings for Steel Sheet Roofing

These are used to increase the lifespan of roofing sheet, with PVC, PVF2, polyester, and acrylic most commonly used.[6] All of these are products of petrochemical processing.

Manufacture

Energy Use
Plastic polymers are produced using high energy processes, with oil or gas as raw materials, which themselves have a high embodied energy.[13] PVC has the lowest embodied energy of the organic coatings reviewed in this report, at between 53MJ kg^{-1}[8] and 68MJ kg^{-1}.[12]

Resource Use (non-bio)
Oil is the main raw material for the organic coatings listed in this report. This is a non-renewable resource. One tonne of PVC requires 8 tonnes of crude oil. This is less than many other polymers because 57% of its weight consists of chlorine derived from salt, giving PVC a resource advantage over other plastics.[7]

Global Warming
Petroleum refining and synthetic polymer manufacture are major sources of NOx, CO_2, methane and other 'greenhouse' gases.[16, 25]

Acid Rain
Petrochemicals refining and synthetic polymer manufacture are major sources of SO_2 and NOx, the gases responsible for acid deposition.[13, 16]

Toxics
The petrochemicals industry is responsible for over half of all emissions of toxics to the environment.[16]

PVC is manufactured from the vinyl chloride monomer and ethylene dichloride, both of which are known carcinogens and powerful irritants.[9,13] PVC powder provided by the chemical manufacturers is a potential health hazard and is reported to be a cause of pneumoconiosis.[7] High levels of dioxins have been found around PVC manufacturing plants[14] and the waste sludge from PVC manufacture going to landfill has been found to contain significant levels of dioxins and other highly toxic compounds.[15] PVC manufacture is top of HMIP list of toxic emissions to water, air and land;[14] emissions to water include sodium hypochlorite and mercury; emissions to air include chlorine and mercury - although the mercury cells, which release mercury, are being phased out.[25] Acrylonitrile, used in the manufacture of acrylic resins, is a suspected carcinogen and is reported to cause headaches, breathing difficulties and nausea.[9,10] Although associated with toxic emissions, polyester has a relatively small impact when compared with PVC.[11]

Use

Health
PVC is relatively inert in use in construction.[25] We found no evidence of a health risk during use from any of the organic coatings reviewed in this report.

Durability
A 1993 BRE survey found PVC and PVF2 to be the most durable organic coatings. Acrylic coating was found to have a high faliure rate (leading to a high incidence of random and cut-edge corrosion of the underlying steel), and the report suggests that acrylic may be an inappropriate covering due to its low durability.[6]

Recycling/Reuse/Disposal
There are concerns about the creation of toxins when organic coatings are burned in an electric arc furnace during steel recycling.[1] For example, PVC can form highly toxic polychlorinated dibenzo-dioxins (dioxins) when burned.[7] It is unlikely that removal of organic coatings prior to recycling will be economically feasible in the foreseeable future.

Other
The extraction, transport and refining of oil to produce organic polymers can have enormous localised environmental impacts.[13]

ALERT
Phthalates used as plasticisers in PVC together with dioxins have been identified as hormone disrupters, with possible links to a reduction in the human sperm count, and disruption of animal reproductive cycles.[43] Hormone disrupters operate by blocking or mimicking the action of particular hormones. Humans are most affected through the food chain, and unborn children absorb the toxins through the placenta, and babies through their mothers milk.[57]

PVF2

We were unable to find ANY information regarding the environmental impacts of PVF2, which do not appear to have been studied by either industry or campaigning groups. It is likely that the main impacts of this fairly inert fluorinated compound will be similar to PVC, and occur during manufacture and incineration/disposal. Fluorine, used in its manufacture, is a particularly aggressive corrosive element.

(b) Stainless Steel Sheet

Manufacture

Energy Use

Stainless steels from recycled scrap (see 'resource use') consume 11MJ kg^{-1}, 20% of which is from collection of scrap and distribution of the product.[17]

Resource Use (non-bio)

According to CIRIA, the principal feedstock for stainless steel in the UK is scrap steel, and stainless steels are not believed to be produced from ore anywhere in the world.[17] We found no data regarding the origins of the nickel, molybdenum, chromium, vanadium or other alloying metals used in stainless steel manufacture.

Global Warming

About 1.6 tonnes of CO_2 is emitted per tonne of stainless steel produced from recycled scrap.[17]

Toxics

Nickel, vanadium, molybdenum and chromium released in scrubber effluents can be toxic and phytoxic (toxic to plants), and are released in this country to within NRA/HMIP consent limits. Mercury and copper may also be found in wastes. Heavy metals, cyanide, carbonyls, oils, fluorides and other toxins may be released to air, within HMIP consent limits.[17]

Also see Steel Sheet 'toxics' section, p.154.

Acid Rain

SO_2 and NOx arise from fuels consumed in production.[17] The smelting of molybdenum and other alloying metals results in the emission of sulphuric acid fumes, which can lead to local problems of acid deposition. Particular problems of sulphuric acid 'spotting' have been experienced around the Glossop molybdenum smelter, due to 'grounding' of the smelters chimney plume.[59]

Use

Durability

Stainless steels have a longer service life than mild steels, with little or no maintenance.[17]

Recycling

Stainless steel is highly recyclable and is only produced by recycling in the UK.[17]

ALERT

See Steel Sheet 'Alert' section, p.172

(c) Aluminium Sheet

Manufacture

Energy Use

Aluminium has an extremely high embodied energy of 180-240MJ kg^{-1}.[17] (or 103,500 Btu at point of use)[5] The aluminium industry accounts for 1.4% of energy consumption worldwide,[5] the principle energy source being electricity. Recycled aluminium gives an 80%-95% energy saving over the virgin resource at 10 to 18 MJ kg^{-1}.[5, 17] The production of aluminium uses energy for the heating of initial bauxite-caustic soda solutions, for the drying of precipitates, for the creation of electrodes which are eaten up in the process, and for the final electrolytic reduction process.[3] Finishing processes such as casting or rolling require further energy input. Bear in mind that most embodied energy figures are quoted on a per tonne or per kilogram basis - which ignores aluminium's low density compared to say steel. Of the four aluminium smelters in the UK, though the two small Scottish plants use hydro-power, the larger plants use coal (Lynemouth) and national grid (nuclear) electricity (Anglesey).[77] It is claimed by some commentators that energy consumption figures for aluminium can be misleading, as the principle energy source for virgin aluminium manufacture is electricity produced from hydroelectric plant and is therefore a renewable resource.[17]

Resource Use (bio)

Bauxite strip mining causes some loss of tropical forest.[5] The flooding of valleys to produce hydroelectric power schemes often results in the loss of tropical forest and wildlife habitat, and the uprooting of large numbers of people.

Resource Use (non-bio)

Bauxite, the ore from which aluminium is derived, comprises 8% of the earths crust.[5] At current rates of consumption, this will serve for 600 years supply, although there are only 80 years of economically exploitable reserves with current market conditions.[17]

Global Warming

The electrolytic smelting of aluminium essentially comprises the reaction of aluminia oxide and carbon (from the electrode) to form aluminium and carbon

Health Effects of Lead

Lead is a cumulative poison. About 90% of the lead retained in the body enters the bones, where it can be stored for 40 - 90 years, leading to accumulation.[71] This may be remobilised during periods of illness, cortisone therapy, and in old age.[8] The effects of lead toxicity are very wide ranging and include impaired blood synthesis, hypertension, hyperactivity and brain damage.[71] Research indicates that lead may blunt the body's immune systems, which implies that a consequence of chronic (long term, low dose) lead poisoning could be an increase in susceptibility to infection and disease.[8] It is suspected that lead is associated with mental retardation in children,[8] and it has been suggested that lead poisoning was partially responsible for the fall of the Roman empire, where large amounts of lead were used for purposes ranging from water pipes to a food sweetener!

dioxide, the greenhouse gas.[64] Globally this CO_2 production is insignificant compared to the contribution from fossil fuel burning, but compared to iron and steel, aluminium produces twice as much CO_2 per tonne of metal (though allowance should perhaps be made for the lower density of aluminium). Nitrous oxide emissions are also associated with aluminium production.[78]

One tonne of aluminium produced consumes energy equivalent to 26 to 37 tonnes of CO_2 - but most imported aluminium is produced by hydroelectric power with very low CO_2 emission consequences.[17]

Acid Rain

SO_2 and NOx are released when fossil fuels are burned at all stages of manufacture, to produce electricity (see 'global warming' above) and in gas-fired furnaces.[17]

Toxics

Bauxite refining yields large volumes of mud containing trace amounts of hazardous materials, including 0.02kg spent 'potliner' (a hazardous waste) for every 1kg aluminium produced.[5]

Fabrication and finishing of aluminium may produce heavy metal sludges and large amounts of waste water requiring treatment to remove toxic chemicals.[5]

Aluminium processes are prescribed for air pollution control in the UK by the Environmental Protection Act 1990,[22] and emissions include hydrogen fluoride, hydrocarbons, nickel, electrode carbon, and volatile organic compounds including isocyanates.[17]

Metal smelting industries are second only to the chemicals industry in terms of total emissions of toxics to the environment.[16]

Aluminium plants in the UK have been frequently criticised for high levels of discharge of toxic heavy metals to sewers.[80]

Emissions of dioxins have also been associated with secondary aluminium smelting.[79]

Other

The open-cast mining of the ore, bauxite, and of the limestone needed for processing can have significant local impact, bauxite mining leaving behind particularly massive spoil heaps.[81]

The association between aluminium smelting and large scale hydro-electric dams in third world countries is well known. So too is the damage such schemes cause to both human communities and to the natural environment.[64,82]

Use

Durability

Aluminium coating and sheet is naturally protected by an impermeable oxide layer which forms on the surface,[36] which makes aluminium more durable than steel sheet.

Recycling/Reuse/Disposal

Aluminium coatings are not usually removed for recycling,[1] although aluminium sheet is readily recycled. Some aluminium roofing products such as Rustic Shingle from Classic Products Inc, Ohio USA, contain up to 98%

recycled metal, which reportedly uses only 15% of the energy of virgin material.[1]

Powder coated aluminium is not recyclable.[46]

(d) Lead Sheet

Manufacture

Energy Use

Total energy for primary production of lead is estimated at 190MJ kg^{-1}. Production from scrap can use as little as 10MJ kg^{-1}, depending on the quality of the scrap.[17]

Resource Use (non-bio)

Lead is often extracted from the same ore as zinc,[17] and so the same impacts apply with regards to extraction.

Global Warming

CO_2 emissions are estimated at 16 tonnes per tonne of lead produced.[17]

Toxics

Lead is toxic and phytoxic[17] and tends to bioaccumulate - see box below for details of toxic effects. Metals may enter scrubber effluents and other process waters. Emissions to air must be controlled by bag filter, but metals may still escape. Principal pollutants arising are lead metal and oxides, zinc, mercury, cadmium, chromium, copper, fluorides, hydrogen sulphide, carbonyles and antimony.[17]

Acid Rain

Emissions of SO_2 and NOx can be substantial due to the fuels consumed in lead manufacture.[17]

Other

As with other metals, the extraction of lead is locally damaging.[46]

Use

Occupational Health

There is a possible health risk from skin contact with lead as it is readily absorbed through the skin.[46]

Recycling/Reuse/Disposal

Lead is highly recyclable, has a high scrap value and is comprehensively recycled.[17]

Health

Lead poisoning is characterised by disruption of the metabolism, destruction of red blood cells, damage to internal organs and neural damage. Lead may enter the body through ingestion, inhalation and skin contact.

Lead roofing elements may contaminate run-off water, making it unsuitable for collection as irrigation or potable water.

(e) Copper Sheet

Manufacture

Energy Use

Energy is consumed in mining, smelting or electrowinning, refining, melting and product fabrication. Embodied energy is estimated at 70MJ kg^{-1}, 9% of which is in transport to the UK.[17] Production from scrap uses between 10 and 60MJ kg^{-1}, depending on purity.[17]

Resource Use (bio)

Clearance for copper mining results in habitat loss.[21] The 0.45mm copper sheet used in roofing is approximately one fifth of the weight of tiles, allowing for savings in timber substrates.[2]

Resurce Use (non-bio)

We found no estimates of remaining exploitable reserves.

Global Warming

About 7 tonnes of CO_2 are produced per tonne of copper produced from ore, and 1-6 tonnes per tonne of recycled copper.[17]

Acid Rain

SO_2 and NOx emissions will be substantial due to the fuels consumed during copper manufacture.[17, 21]

Toxics

Heavy metals are often leached into watercourses from mine drainage and spoil tips, with associated acidification of water (Acid Mine Drainage).[21] For example, the Afon Goch ('Red River') on Anglesey is heavily polluted with heavy metals and has a pH as low as 2, due to leachate from the Parys Mountain copper mine.

Copper mining also yields large amounts of heavy metal contaminated solid waste, and emissions to air include heavy metals, carbonyls, fluorides, alkali and acid fumes, dust and resin fume.[17]

In the UK, copper processing is a prescribed process, and emissions are controlled by HMIP/NRA (now the Environment Agency) consent levels.[21, 24]

Other

There are likely to be localised impacts of vibration, noise and dust created during ore extraction.[21]

Use

Durability

Copper has a high degree of corrosion resistance[2] and, unlike steel, can be used without protective coatings. Copper roofing is reported to last up to 100 years[(2)] - a lifetime similar to slate.

Recycling/Reuse/Disposal

Copper is recyclable,[2] and 60-70% is estimated to be recycled.[17]

12.5 Alternative Roofing Materials

12.5.1 Wooden Shakes & Shingles

A common roofing material in the USA are shakes and shingles, made from split (shakes) or sawn (shingles) wood, typically cedar. It is claimed that cedar shingles are virtually indestructible, and can be bent around gentle curves, thus suited to contoured roofs. They are also light, requiring less supporting structure than many conventional roofing products.[53]

While wood shingles are initially seen as an environmentally benign roofing material, being derived from a potentially renewable resource, requiring very little energy in manufacture, having a relatively high insulation value and being both biodegradable and durable,[1,53] they have a few potential drawbacks. In the USA, Cedar, the favoured wood for shingles, is currently unsustainably cut from old growth forest, and there has been no concerted effort to replant cedar forests.[1] In tropical countries the intensive use of wood as a roofing material contributes to rainforest destruction,[36] although some manufacturers claim that their shingles are cut from 'waste' material which is unsuitable for other products, and many suppliers source their cedar from Canada, where forestry practices are reported to be more environmentally sensitive.[53,86] Plantation grown Western Red Cedar may be available (although not certified to date) and it is reported to grow well in the UK.[76]

In the USA, shingles are reported to be 'notorious fire hazard',[1] although this is not thought to be a problem in the wetter climate of the UK. Some treatments used for fireproofing in the US have been found to be carcinogenic.[1], although we are assured by manufacturers that this is not the case in the UK. Wood shingles are reported to require frequent maintenance to prevent dirt from building up between the shingles and causing rot,[1] although architect Brian Ford, who used fire retardant Cedar shingles on the new School of Engineering building at the DeMontford University, Leicester, claims that shingles were chosen for their low maintenance and long life.[54] On balance, we would consider wood shingles or shakes as the best environmental option if made from a certified sustainable source, using 'safe' fireproofing and preservatives, with an owner who is prepared for the possibility of a high level of maintenance!

CSA certification comes on stream this year, and Silva certification is currently seeking accreditation to the FSC.[76] Information on certified sustainable timber sources can be found in Chapter 7.

12.5.2 Used Tyres

Moore Enviro Systems and Tread Mill Inc. in the USA use the tread of used tyres as roofing tiles. At a fitted cost of between $1.50 and $2.00 per square foot (around £10 to £14m^{-2}), this appears to be both an environmentally and financially attractive option - although aesthetic and fire concerns will have to be overcome.[1] We were unable to find any companies offering a similar service in the UK.

12.5.3 Recycled PVC & Wood Shingles

'Eco-shake' shingles, manufactured from 100% recycled wood and PVC, have recently entered the market after 'intensive testing'.[75] The shakes exceed all the relevant US standards, achieve a class-A fire rating and have a 50 year warranty.[75]

The shingles are reported to cut and nail just like wood, and are aesthetically an ideal substitute for wood - available in three colours, and with heavily textured wood grain to replicate weathered cedar.

Although made entirely from reprocessed materials, these are post-industrial rather than post consumer. Wood fibre is obtained from sawmills and cabinet makers, and reclaimed PVC from garden hose manufacturing scrap.[75] For more information:

Re-New Wood, Inc.
P.O. Box 1093
Wagoner, OK 74467
USA
800/485-5803,
918/485-5803

12.5.4 Thatching

Historically speaking, thatch is the 'original' roofing material. Although with the exception of rural Asia and Africa, it has been largely replaced by rigid stone, clay, cement or metal sheet roofing, thatch is still used in NW Europe because it is weathertight, durable and aesthetically pleasing.[35]

Environmentally, thatch has the advantage of making use of a renewable and local resource. Harvesting of materials does not require sophisticated tools,[35,36] and the reeds used for thatching in the UK grow back annually.[35] Thatch provides excellent insulation, a 300mm thatch providing the same insulation as 200mm fibre-glass matting under corrugated iron.[35]

Thatch is often not thought of as being durable, and many traditional types of thatch have very real shortcomings - they leak, harbour insects, tend to catch fire, and do not last long.[35] Installation is very labour intensive, making thatching expensive in the UK where labour costs are high. However, with modern techniques, thatching with certain types of grass can last 50 years or more if carefully laid,[35] and are therefore comparable in durability to metal sheet, fibre-cement, concrete, asphalt and the cheaper clay tiles. Nevertheless, thatch is unsuitable for high density housing because it is combustible, although measures can be taken to reduce the fire risk.[35]

In the UK, the most important thatching material is the reed

Phragmites australis. Reeds are much stronger than straw, and are therefore the preferred material for thatching. Norfolk reed is the best in the UK, growing to a height of 2.5m, and highly durable, lasting for 50-60 years in a roof thatch.[34] The presence of nitrates weakens the durability of reeds, and so organic sources, unaffected by agricultural run-off, need to be used for thatching.[34]

12.5.5 Bamboo

Bamboo is used as a roofing material in tropical countries using two main techniques; Bamboo tiles consisting of successive rows of half stems covering the whole length from ridge to gutter, the first layer with the concave side facing up, the second with the concave side facing down, as for Roman tiles. Bamboo shingles are segments of 3-4cm in width, and maximum length between knots. These are fixed on battens made from half stems, laid with the concave side up.[36] The advantage of bamboo is that it grows easily, rapidly and in abundance. It also has excellent bending and tensile characteristics. The main drawbacks are its high flammability, low moisture resistance and short lifespan.[36]

12.5.6 Plastic Panels

The benefits of this material depend on whether the plastic is recycled or not.

There have been reports from the USA of re-ground plastic from computer casings being used as roofing tiles,[4] although we have been unable to track down a supplier or obtain details of performance.

The president of Nailite, a company which briefly manufactured plastic roofing on a prototype basis, stated that "I am not confident that plastic, especially painted plastic, can be depended on for more than ten years".[1] Considering that plastic is manufactured entirely from a limited non-renewable resource (oil), and requires large amounts of energy in its manufacture, it does not seem a particularly wise environmental option considering its low durability - unless the plastic is recycled.

12.5.7 Planted Roofs

Covering a roof with vegetation can create an attractive finish, providing additional areas of greenery in cities, or helping to reduce the visual impact of rural buildings. Erisco-Bauder Ltd of Ipswich (and other manufacturers, listed on p.162), offer systems which allow architects to specify vegetated roofs as easily as they would an ordinary flat roofing system.[45] Two types of roof are offered: the equivalent of a 'normal' garden, with trees and plants, needing intensive maintenance similar to a normal garden, and an 'extensive' system which is light in weight and uses slow growing low or no maintenance plants.[44]

It is also possible to design your own roof garden system. A planted roof to take a lawn and small shrubs will have a soil depth of 100mm, weighing up to 1.15kN m^{-2}; For large bushes and trees, a soil depth of 350mm will be

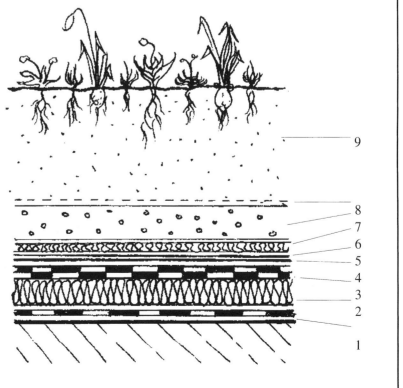

Typical Planted Roof

1. Vapour barrier bonded to roof deck.
2. Insulation, which is able to withstand water and pressure of the soil.
3. Waterproofing and vapour equalization layer.
4. Root barrier and second waterproof layer.
5. Separation layer, to allow relative movement between the planted layer and the waterproofing below.
6. Protection layer to prevent damage to the layers below.
7. Drainage or water retention system to prevent waterlogging/drying out.
8. Filter layer to prevent soil particles blocking the drainage.
9. Soil and planting.[44]

required, with a loading of 4.0kNm^{-2}. Pre-planted roofs with small meadow flowers and grasses or turfed roofs are also available, with loadings of 0.8 and 0.65kN m^{-2} respectively - comparable to that of concrete tiles with underfelt and battens.

A planted roof requires a fairly complex system to prevent problems of root penetration and water seepage (see diagram below).

It is argued that a planted roof is environmentally preferable to an asphalt roof because it creates space for plants and insects, provides food for birds, and even a habitat for small mammals if the roof is large enough,[44] as well as being aesthetically pleasing. However, it must be remembered that due to the additional weight of the soil and the extra protective layers required in the roof structure, green roofs can be far more materials intensive than a conventional flat roof.

The impacts of materials used in planted roofs are discussed in issue 14 of the Green Building Digest magazine.

12.5.8 Photovoltaic Roofing Panels

The large unused area available on buildings in urban locations offers the potential for significant point-of-source electricity generation in the UK.[50] It is estimated that the total energy resource from photovoltaic cladding and roofing systems in the UK is equivalent to 68GW generating capacity, with 1995 technology.[51]

The use of photovoltaic (PV) roofing elements is particularly attractive for commercial buildings, as their main energy demand is during the day, when the resource (the sun) is available.[50, 51] The use of PV roofing systems can also provide an 'environmental statement' for the building owner.

PV systems are currently considered uneconomic in terms of the unit cost of electricity produced,[50] except in areas of high sunlight such as southern America, North Africa and southern Europe.[51] However, the offset costs of a conventional roofing system should also be taken into account and furthermore, heat recovery from the back of the PV panel has been found to increase generation efficiency as well as providing potential for passive space and water heating. If the currently 'externalised' costs of repairing damage caused by conventional power generation are considered, solar generation is arguably economic already.[50] There are encouraging forecasts which suggest that PV cladding systems could reach cost competitiveness in the UK in terms of unit generating costs by the year 2008.[50, 52] It may be worthwhile designing roofs to have the correct southfacing slope/aspect for solar panels, in anticipation that when re-covering is needed the technology may well be much better and cheaper.

The main environmental impact of PV technology is the high embodied energy of the generating panels, which it is estimated may be 'paid back' within 5 years in the UK, and within a few months in sunnier climates. Considering the typical lifetime of a PV module is 20-30 years, the energy payback is at least four times the embodied energy, and may be repaid up to 60-100 times over.[50]

Impacts which are harder to reconcile are the production of toxins such as diborane, silica dust, cadmium compounds and gallium arsenide during manufacture. Research into the recycling potential of PV cells may help to reduce this impact.[50]

Solar Roof

The Centre for Alternative Technology (CAT) at Machynlleth have completed the UKs first residential Solar Study Centre, which features the "largest complete roof on a normal building to be constructed entirely in photovoltaic material".[56] The roof is expected to generate 10,000 kWh of electricity per year - enough to meet the needs of three houses. In order to maximise the use of the energy generated, surplus energy is exported to the national grid using a "Powergate" system.[56]

Contact:
CAT, Machynlleth, Powys, Wales SY20 9AZ. Tel. 01654 702400

Passive Solar

Passive solar systems, used to directly heat air or water, can also be used as roofing elements. Due to time considerations, we were unable to cover these - but details of these systems can be obtained from CAT at the address above.

Specialist Suppliers

Planted Roofs

GRODAN Sloping roof garden system (lawn slab) & Extensive Roof Garden System

Grodania A/S, Wern Tarw Pencoed, Bridgennd Mid Glamorgan, CF35 6NY, Wales (tel. 01656 863853 fax. 01656 863611)

A Swiss manufactured product, consisting a turf over a Savanna slab.

Embodied energy: 14 MJ/kg

Also produce lawn slabs for intensive or extensive applications

Embodied enerfy: 230 MJ/kg

EUROROOF Ltd, Denton Drive, Northwich, Cheshire, CW9 7LU (tel. 01606 48222 fax. 01606 49940)

Intensive and extensive roof garden systems, with two alternatives available for those wishing to avoid specifying WSB 80 PVC membrane.

ERISCO BAUDER Ltd, Ipswich, (tel. 01473 257671)

Natural Slate

DELABOLE SLATE Co. Ltd, Pengelly Road, Delabole, Cornwall, PL33 9AZ (tel. 01840 212242 fax. 01840 212948)

Embodied Energy: 540 MJ/kg

CWT-Y-BUGAIL SLATE Co. Ltd, Plas y Bryn, Wynne Road, Blaenau, Festiniog, Gwyned, LL41 3RG (tel. 01766 830204 fax. 01766 831105)

Produce natural Welsh roofing slates, certified to BS 5534 Pts 1 & 2 and BS 680 Pt 2, (BSI Registered) for roof pitches as low as 20 degrees.

Enbodied Energy: 540 MJ/kg

GLODDFA GANOL SLATE MINE, Blaenau, Ffestiniog, Gwynedd, LL41 3NB, Wales (tel. 01766 830664 fax. 01766 830527)

Embodied Energy: 540 MJ/kg

GREAVES WELSH SLATE Co. Llecheedd Slate Mines, Blaenau, Ffestiniog, Gwynedd LL41 3NB (tel. 01766 830522 fax. 01766 831064)

Embodied Energy: 540 MJ/kg

Sheet Metal

'TECU' Copper Sheet & Fittings

KM - KABELMETAL UK Ltd, 9-17 Tuxford Road, Hamilton Industrial Estate, Leicester, LE4 7TZ, England (tel. 01162 461130 fax. 01162 461132)

Embodied Energy: 133,000 MJ/kg

Copper roofing & Cladding

OUTOKUMPU (UK) Ltd, Outokumpu House, 10 Hammersmith Broadway, Greater London W6 7AL (tel. 0181 741 3141 fax. 0181 741 7164)

Embodied Energy: 133,000 MJ/kg

CALDER INDUSTRIAL MATERIALS Ltd, Crescent House, Newcastle Upon Tyne, Tyne & Wear NE99 1GE (tel. 0191 261 0161 fax. 0191 261 1001)

Produce a wide range of sheet and flashing for those wishing to avoid lead - using Copper, Aluminium, Titanium-Zinc etc.

'UGINOX' Stainless Steel Flashing

EUROCOM ENTERPRISES Ltd, Index House, St Georges Lane, Ascott, Berkshire, SL5 7EU (tel. 01344 23404 fax. 01344 874696)

Contact: Francois Moal

Produce a lead free stainless steel material that has been 'terned' with tin to provide the same appearance as lead where planning requirements dictate. The material can be used for flashing or total roof covering. Useful where a totally inert material in required

Embodied energy: 103,000 MJ/kg

BILLITON METALS UK Ltd, c/o Nedzinc, 84 Fenchurch Street, London EC3M 4BY (tel. 0171 860 8280 fax. 0171 481 0017)

Produce a Zinc Titanium Roofing and flashing system for those wishing to avoid lead sheet and flashings.

Handmade Clay Tiles

KEYMER BRICK & TILE CO. Ltd, Nye Road, Burgess Hill, West Sussex, RH15 0LZ (tel. 01444 232931 fax. 01444 871852)

Embodied Energy: 1520 MJ/kg

TUDOR ROOF TILE Co. Ltd, Denge Marsh Road, Lydd, Kent, TN29 9JH (tel. 01797 320202 fax. 01797 320700)

Produce range of hand crafted clay tiles, nibbed or for pegs in two standard colour ranges

Embodied Energy: 1520 MJ/kg

Clay Roof Tiles

HAWKINS' TILES, Watling Steet, Cannock, Staffordshire, WS11 3BJ (tel. 01543 502744 fax. 01543 466434)

HILTON PERRY & DAVENHILL Ltd, Dreadnought Works, Dreadnought Road, Pensnett, Brierley Hill, West Midlands, DY5 4TH (tel. 01384 77405 fax. 01384 74553)

Embodied Energy: 1520 MJ/kg

Wooden Shingles & Shakes

'COLT PRECEDA', THE LOFT SHOP LTD, Horsham Gates, North Street, Horsham, West Sussex RH13 5PJ (tel. 01403 274275 fax. 01403 276277)

Contact: Marjory Kay (Environmental Officer)

Cedar and Sweet Chestnut shingles and shakes produced in the USA. treated with water borne preservative to BS 4072: 1974. 400mm length, width varying between 75-355mm. Shingles taper from tip to butt (2-10mm)

Embodied Energy: 7540 MJ/kg

EWAN CLITHEROW-BARKER Quarter Sawn Oak Roof Shingles, Dolen Farm ISAF, Eglwysfach, Nr Machynlleth, Powys SY20 8SX, Wales (tel. 01654 781354)

Embodied Energy: 220 MJ/kg

JOHN BRASH & Co Ltd Tapered Cedar Shingles, The Old Shipyard, Gainsborough, Lincolnshire DN21 1NG (tel. 01427 613858 fax. 01427 810218)

Western Red Cedar hand split shingles for vertical hanging or roof pitches over 20 degrees. Produced in the USA.

Embodied Energy: 7540 MJ/kg

Roofing Membranes

(See Green Buuilding Digest Magazine, Issue 14 for Impact Analysis)

FIRESTONE BUILDING PRODUCTS Ltd Rubberguard EPDM Roofing Membrane, Strayside House, West Park, Harrogate, West Yorkshire, HG1 1BJ (tel. 01423 520878 fax. 01423 520879)

Produce EPDM single ply roofing membrane with a weight less than 1.4kg/m2 and a thickness of 1.15mm. Other thicknesses are available.

I.C.B. Ltd (ALWITRA) EDPM Single Ply Flat Roof Membrane,, Unit 9, West Howe Industrial Estate, Elliot Road, Bournemouth, Dorset BH11 8JX (tel. 01202 579208 fax. 01202 581748)

Contact:Bob Dixon

Evlastic S and SV are an EPDM rubber dispersion in a polyolefine matrix. Suitable for fully bonded or loose-laid and ballasted finish. Approved for use in contact with or for containment of potable water by authorities in the USA, Belgium and Germany.

Embodied Energy: 47,000 MJ/kg

Reclaimed Products

A comprehensive and up to date listing of suppliers specialising in reclaimed roofing products is produced by SALVO,
tel. (01668) 216494.

(1) Environmental Building News 4 (4) July/August 1995

(2) Copper in Roofing, Cabling and Plumbing (B. Findlay). In: The European Directory of Sustainable and Energy Efficient Building. O. Lewis & J Goulding, Eds. James & James Science Publishers Ltd 1996

(3) Construction Materials Reference Book (D.K. Doran, Ed). Butterworth Heinmann 1992

(4) "Designs for Disposal" Compressed Air **97** (8) p.19-23 December 1992

(5) The Greening of the Whitehouse. http://solstice.crest.org/environment/gotwh/general/materials.html

(6) Survey of Performance of Organic-Coated Metal Roof Sheeting. (R.N. Cox, J.A. Kempster & R. Bassi). Building Research Establishment Report BR259. 1993

(7) Greenpeace Germany Recycling Report, 1992

(8) Dictionary of Environmental Science & Technology (A. Porteous) John Wiley & Sons. 1992

(9) H is for ecoHome (A. Kruger). Gaia Books Ltd, London. 1991

(10) The Non-Toxic Home. (D.L. Dadd). Jeremy P. Tarcher Inc, Los Angeles. 1986

(11) Production & Polymerisation of Organic Monomers. IPR 4/6. Her Majesties Inspectorate of Pollution, HMSO 1993

(12) The Consumers Good Chemical Guide. (J. Emsley) W.H. Freeman & Co. Ltd, London. 1994

(13) Green Building Digest no. 5, August 1995

(14) PVC: Toxic Waste in Disguise (S. Leubscher, Ed.). Greenpeace International, Amsterdam. 1992.

(15) Achieving Zero Dioxin - an emergency stratagy for dioxin elimination. Greenpeace Interational, London. 1994.

(16) The Global Environment 1972-1992; Two Decades of Challenge. (M.K. Tolba & O.A. El-Kholy (Eds)) Chapman & Hall, London, for the United Nations Environment Program. 1992

(17) Environmental Effects of Building and Construction Materials. Volume C : Metals. (N. Howard). Construction Industry Research and Information Association, June 1995

(18) Metal Statistics 1981-1991, 69th Edition. Metallgesellschaft AG. Frankfurt-am main 1992

(19) UK Iron and Steel Industry: Annual Statistics. (UK Iron & Steel Statistics Bureau). 1991

(20) Pollution - Causes, Effects and Control. 2nd Edition (R.M. Harrison). Royal Society of Chemistry. 1990.

(21) Environmental Aspects of Selected Non-ferrous Metals Ore Mining - A Technical guide. Unired Nations Environment Programme Industry and Environment Programme Activity Centre. 1991

(22) Environmental Protection Act 1990 Part I. Processes prescribed for air pollution control by Local Authorities, Secretary of State's Guidance - Aluminium and aluminium alloy processes. PG2/6(91). (Department of the Enviroment) HMSO 1991

(23) Environmental Protection Act 1990 Part I. Processes prescribed for air pollution control by Local Authorities, Secretary of State's Guidance - Iron, steel and non-ferrous metal foundry processes. PG2/4(91). (Department of the Enviroment) HMSO 1991

(24) Environmental Protection Act 1990 Part I. Processes prescribed for air pollution control by Local Authorities, Secretary of State's Guidance - Copper and copper alloy processes. PG2/8(91). (Department of the Enviroment) HMSO 1991

(25) Environmental Impact of Building and Construction Materials. Volume D: Plastics and Elastomers (R. Clough & R. Martyn). June 1995

(26) ENDS Report 245, p.27-29 June 1995

(27) ENDS Report 244, p.22-25 May 1995

(28) Hazardous Waste and Cement Kilns. Friends of the Earth Briefing Sheet, June 1995

(29) Incineration by the Back Door - Cement Kilns as Waste Sinks. (T. Hellberg). The Ecologist, 25 (6) p.232-237 November/December 1995

(30) Environmental Building News 4 (5) p.5 September/October 1995

(31) Mitchells Materials, 5th Edition (A. Everett & C.Barritt) Longman Scientific & Technical, England. 1994

(32) Personal Communication, Thornton Kay, SALVO, March 1996

(33) Construction Materials Reference Book (D.K. Dorran) Butterworth Heinmann, Oxford. 1992

(34) Eco-renovation - the ecological home improvement guide. (E. Harland). Green Books Ltd, Devon. 1993

(35) Thatching: A Handbook (N. Hall). Intermediate Technology Publications, London. 1988.

(36) Fibre and Micro-Concrete Roofing Tiles - Production process and tile laying techniques. International Labour Organisation Technology Series, technical memorandum no.16, 1992.

(37) Environmental Impacts of Building and Construction Materials Volume B: Mineral Products. (R. Clough & R. Martin). Construction Industry Research and Information Association, London. June 1995

(38) Environmental Protection (Prescribed Processes and Substances) Reguations 1991 No 472. Department of the Environment. 1991

(39) ENDS Report 243, p.23-24. April 1995

(40) Admixture Data Sheets, 9th Edition. Cement Admixtures Association. 1992

(41) Environmental Issues in Construction (SP94). Construction Industry Research and Information Association, London 1993

(42) Technical Note on Best Available Technology Not Entailing Excessive Cost for the Manufacture of Cement. (EUR 13005 EN) Commission for the European Community, Brussels 1990

(43) Greenpeace Business No.30 p5, April/May 1996

(44) Green Architecture (B Vale & R Vale) Thames & Hudson Ltd, London. 1991

(45) Erisco-Bauder Green Roof Systems. Erisco-Bauder Ltd, Ipswich, Suffolk, 1989

(46) The Green Construction Handbook. (Ove Arup & Partners). JT Design Build Publications, Bristol. 1993

(47) Asbestos Materials in Buildings. (Department of the Environment) HMSO, 1991.

(48) Asbestos in Housing (Department of the Environment) HMSO, 1986

(49) C is for Chemicals - Chemical Hazards and How to Avoid Them. (M. Birkin & B. Price) Green Print/Merlin Press Ltd, London. 1989.

(50) A Feasibility Study of Passive Combined Heat and Power Systems and their Potential for Integration into UK Building Facades. (Orrock, Price, Satharasinghe & Stradian) MSc Thesis, Energy Systems and the Environment, University of Strathclyde. 1995

(51) Photovoltaics in Buildings - Costs and Economics (N.M. Pearsall) Newcastle Photovolatics Applications Centre, University of Northumbria. September 1995

(52) The Use of PV Cladding for Commercial Buildings in the UK: Costs and Conclusions. (F. Crick, J. Louineau, B. McNelis, R. Scott, B. Lord, R. Noble, & D. Anderson) Paper presented at the 12th European Photovoltaic Solar Energy Conference, Amsterdam, April 1994.

(53) 'Environmentally Friendly Centre' Chooses Shingles for Roof. Press Release, the Loft Shop. June 1995

54. Cedar Shingles Add the Finishing Touch to DeMontfort University, Leicester. Press Release, The Loft Shop. January 1994.

55. Ferrocement Roofing Element. (K. Sashi Kumar, P.C. Sharma & P. Nimityongskul) International Ferrocement Information Centre (IFIC) Do It Yourselt Series, Booklet No. 5, 1985

56. Help us Build the Roof of the Future. Centre for Alternative Technology Solar Roof Appeal leaflet and letter. Undated (Recieved April 1996).

57. Taking Back Our Stolen Future - Hormone Disruption and PVC Plastic. Greenpeace International. April 1996

58. Personal Communication, Thornton Kay at SALVO, April 1996

(59) Personal Communication, Dr Chris Woods, Dept. of Planning, the University of Manchester, Spring 1995.

60. Fax Message, Eternit UK Technical. 16th April 1996

61. Spons Architects and Builders Price Book. (Davis, Langdon & Everest) E & F.N Spon, London. 1995

62. Buildings and Health - the Rosehaugh Guide (Curwell, March & Venables) RIBA Publications 1990

63. Minerals Planning Guidance: Guidelines for Aggregate Provision in England. DoE 1994.

64. Greener Buildings Products and Services Directory. (K. Hall & P. Warm). Association for Environmentally Conscious Building.

65. The Penguin Dictionary of Building, 4th Edition (J. Maclean & J. Scott). Penguin Books, 1993

66. Glass Fibres and Non-Asbestos Mineral Fibres. IPR 3/4 (Her Majesties Inspectorate of Pollution), HMSO London, 1992

67. Rachels Hazardous Waste News #367 (Environmental Reseaech Foundation, Anapolis MD, USA) December 9th 1993

68. ENDS Report No.240, Jan 1995

69. Dioxins in the Environment, Pollution Paper 27. (Department of the Environment) HMSO London, 1989

70. Chlorine-Free Vol. 3 (1). Greenpeace International, 1994

71. Environmental Chemistry, 2nd Edition. (P. O'Neill). Chapman & Hall, London. 1993

72. Roofing and Cladding in Steel - Product Selector. British Steel Corporation, Feb 1987

73. Colorclad Roof Sheeting Profile R38 & R60. Briggs Amasco, July 1995

74. Santoft Roof Tiles (L238005). Santoft Ltd, Oct 1995

75. Environmental Building News, 5 (3) May/June 1996

76. Personal Communication, Rod Nelson, Woodmark, May 1996

77. UK Minerals Yearbook 1989. (British Geological Survey Keyworth, Nottingham) 1990.

78. Metal Industry Sector IPR 2 (Her Majesties Inspectorate of Pollution). HMSO London. 1991

79. ENDS Report No. 240. January 1995

80. The Secret Polluters (Friends of the Earth UK). July 1992

81. Ecological Building Factpack. (Peacock & Gaylard). Tangent Design Books, Leicester. 1992

82. Environmental News Digest vol.10 no.1 (Friends of the Earth Malaysia). January 1992.

83. Environmental Chemistry, 2nd Edition. (P. O'Neill.) Chapman & Hall, London 1993

84. The Green Construction Handbook - A Manual for Clients and Construction Professionals. JT Design Build, Bristol. 1993.

85. ENDS Report 230, March 1994.

86. Personal Communication, Peter Steadman, the Loft Shop. January 1997

Rainwater Goods 13

13.1 Scope of this Chapter

This chapter looks at the main options available in the market for rainwater goods (i.e. gutters, drain pipes etc.) used to collect rainwater run-off from roofs and channel it to sewers, soakaways or storage. The report does not look in depth at the different paints or other surface finishes that some materials may require, but has tried to take these into account in the Best Buy recommendations. (Paints for materials other than joinery will be covered in a forthcoming issue of the Green Building Digest magazine.)

13.2 Introduction

Polyvinyl-chloride (PVC) is the predominant material used for the manufacture of rainwater goods, primarily because of its low price - but its environmental costs are high. Metals are the main alternatives, but metal smelting industries are second only to the chemicals industry in terms of total emissions of toxics to the environment,[5] and may even be worse in some aspects. The closeness of the ratings for the various materials on the Product Table (overleaf) means that it is hard to choose between them purely on their estimated environmental impact.

Dioxins

Greenpeace and other environmental campaign groups are currently campaigning for an end to the use of chlorine in industrial processes. This includes demanding an end to the production of PVC. Their chief concern is with the potential long-term damage to life on earth by the toxic substances known as 'dioxins' - which are found everywhere that chlorine-based industrial processes and materials are used. Tiny amounts of dioxin can cause cancer, birth defects and damage to the immune system. According to the environmental campaigner Dr. Barry Commoner, *"Dioxin and dioxin-like substances represent the most perilous chemical threat to the health and biological integrity of human beings and the environment."*

There are as yet no reliable figures for overall dioxin emissions from its various sources, but recent estimates from the UK government indicate that one of the major sources of dioxin emissions may in fact be the metal smelting industry.[33] Waste incineration, coal burning and diesel combustion are also major sources. German experience shows that dioxin emissions from smelting and incineration can be significantly reduced - by eliminating the use of chlorinated cutting oils, by attention to combustion controls and abatement techniques, or even by using alternative processes.[33]

Plastics fight back

Not surprisingly, the plastics industry has responded to the environmentalist's attack on PVC. Their arguments are based on three main points - that both metal smelting and diesel combustion are responsible for greater dioxin emissions than PVC production, that PVC is a low-energy plastic, and that it has proved safe and durable in a wide variety of uses. But none of these points convince the environmentalists of the overall benefit of PVC.

Whilst PVC production is (probably) responsible for less dioxin emissions than steel, PVC continues to present a danger if it is finally incinerated at the end of its life.

Embodied energy is only one of a number of factors in assessing the environmental impact of a product, and whilst PVC is one of the lowest energy plastics, the other

---

Water, Water, Everywhere, And Not a Drop To Drink!

When considering the design of a rainwater system, it is worth remembering that water is a resource that tends to get overlooked in the UK, perhaps because of the high rainfall and the lack of water metering. But clean fresh water is getting scarcer, especially in the more highly populated areas of the south east, and it won't be long before metering becomes more widespread. Any truly green building will probably include some degree of rainwater storage and use - whether just for watering the grounds, or for cleaning and flushing, or for all fresh water needs.

factors are felt by environmentalists to outweigh such benefits. And, because of the greater problems in recycling PVC over other plastics, this energy advantage may prove less marked in the longer term.

The safety issue concerning PVC is a complex one. In many applications PVC does appear to be safe (the industry quotes its long use as the packaging material for mineral water). Yet in others, there is evidence of harmful chemicals being introduced into people's homes, for example from floor coverings.[28] Ongoing technical developments may help to overcome some of these safety worries, such as with the release of monomers from the plastic. But again this does not address environmentalists key objections, which are concerned with the manufacture and disposal rather than the use of PVC.

The Politics of PVC

As the debate 'hots up', both sides can claim governments and agencies as supporters of their cause. A Swedish government commission recommended that PVC be phased out by the year 2000, and that recycling with current technology could not be recommended. At the same time,

a German government commission supported the continuing use of PVC. Lobbying from both sides no doubt played a large part in both of these conclusions.

The campaign against PVC, and the wider campaign against the industrial use of chlorine in general, is an ambitious attempt to 'clean up' the world's chemical industry. And so the question as to whether or not to choose PVC for any particular application is becoming as much a political as a technical one: do we believe the chemicals industry, or do we believe the environmental campaigners?

Ways Forward

By our rating, all the mainstream alternatives to PVC rainwater goods have probably as high an environmental impact, and yet they all cost a fair amount more. There is an obvious need here for new materials to be developed (or old ones revived) that can fulfil the requirements for cheapness, light weight, ease of use, and environmental acceptability.

The Alternatives section on page 176 mentions one or two ways of doing things differently that might help avoid this dilemma.

Embodied Energy

Embodied energy is the term used to describe the total amount of energy used in the raw materials and manufacture of a given quantity of a product. The first column on the Product Table is an indication of relative embodied energies. As noted elsewhere, in the absence of information on other aspects of a product's environmental impact, embodied energy is often taken to be an indicator of the total environmental impact, and is one of the few quantifiable indicators. The graph below indicates the wide range of values in the literature covering the embodied energy of materials. We have shown the range of values in this case to indicate the difficulty of accurately calculating a figure such as this. Normally, embodied energy figures are given as gigajoules per tonne or per cubic metre, but neither of these is useful here, as different materials are used in different thicknesses. We have adjusted the embodied energy figures to represent the energy in one metre length of 100mm half-round guttering, with thicknesses of 1mm for steel and aluminium, 2.5mm for uPVC and cast iron, and GRP.

Legend:
♦ C.A.T. (29)
□ Roaf & Hancock (30)
– CIBSE (31)
○ Porteous (32)
—▲— Average

Notes:
PVC: figure for 'plastic' used when not specified
GRP: figures for 30% glass and 70% plastic used when not specified
Cast Iron: figure for 2/3 steel used when not specified

	£	Production									Use					ALERT!
	Unit Price Multiplier	Energy Use	Resource Depletion (bio)	Resource Depletion (non-bio)	Global Warming	Ozone Depletion	Toxics	Acid Rain	Photochemical Oxidants	Other	Energy Use	Durability/Maintenance	Recycling/Reuse/Disposal	Health	Other	ALERT!
Rainwater Goods																
Aluminium	4.7	●	·	●			●	·	·	●			●			
Cast Iron	4.3	●	·	●	●		●	●	●	·		·				
Steel	9.5	·	·	●	●		●	●	●	·		●				
Glass Reinforced Polyester	n/a	·		●			·	●	●	●			●			
uPVC	1.0	·		●			●	●	●	·			●	·		Greenpeace Campaign

13.3 Best Buys

Best Buy:	Glass Reinforced Polyester
	Cast Iron

There is no single material that stands out on the Product Table as being environmentally that much better than any other. Given its low price compared to the other options, therefore, some might feel the PVC is the best one to go for.

For those who wish to support the Greenpeace campaign, and for those for whom the extra environmental cost outweighs the monetary cost, the material with the lowest immediate environmental impact would appear from this analysis to be GRP. This is the only material not associated with dioxins, and is certainly worth looking at as a Best Buy for rainwater goods, especially as it is a 'low maintenance' option.

Given the closeness of the scores, some might prefer to choose on the basis of the longevity and final reusability or recyclability of the product. This favours cast iron, which is consequently also a Green Building Handbook Best Buy for rainwater goods.

13.4 Product Analysis

(a) Aluminium

Production

Energy Use

Aluminium is infamous for its high embodied energy, estimated at 180-240MJ kg^{-1}.[42] (or 103,500 Btu at point of use)[43]. The production of aluminium uses energy for the heating of initial bauxite-caustic soda solutions, for the drying of precipitates, for the creation of electrodes which are eaten up in the process, and for the final electrolytic reduction process.[4] Finishing processes such as casting or rolling require further energy input. Bear in mind that most embodied energy figures are quoted on a per tonne or per kilogram basis - which ignores aluminium's low density compared to say steel. (See Embodied Energy below.) Of the four aluminium smelters in the UK, though the two small Scottish plants use hydro-power, the larger plants use coal (Lynemouth) and national grid (nuclear) electricity (Anglesey).[25]

The aluminium industry accounts for 1.4% of energy consumption worldwide,[43] the principle energy source being electricity. Recycled aluminium gives an 80%-95% energy saving over the virgin resource at 10 to 18 MJ kg^{-1}.[42,43] It is claimed by some commentators that energy consumption figures for aluminium can be misleading, as the principle energy source for imported virgin aluminium manufacture is electricity produced from hydroelectric plant and is therefore a renewable resource.[42]

Resource Use (bio)

Bauxite strip mining causes some loss of tropical forest.[43] The flooding of valleys to produce hydroelectric power schemes often results in the loss of tropical forest and wildlife habitat, and the uprooting of large numbers of people.

Resource Depletion (non-bio)

Bauxite, the ore from which aluminium is derived, comprises 8% of the earths crust.[43] At current rates of consumption, this will serve for 600 years supply, although there are only 80 years of economically exploitable reserves with current market conditions.[42]

Global Warming

The electrolytic smelting of aluminium essentially comprises the reaction of aluminia oxide and carbon (from the electrode) to form aluminium and carbon dioxide, the greenhouse gas.[8] Globally this CO_2 production is insignificant compared to the contribution from fossil fuel burning, but compared to iron and steel, aluminium produces twice as much CO_2 per tonne of metal (though allowance should perhaps be made for the lower density of aluminium). Nitrous oxide emissions are also associated with aluminium production.[27]

One tonne of aluminium produced consumes energy equivalent to 26 to 37 tonnes of CO_2 - but most imported aluminium is produced by hydroelectric power with very low CO_2 emission consequences.[42]

Toxics

Aluminium processes are prescribed for air pollution control in the UK by the Environmental Protection Act 1990,[44] and emissions include hydrogen fluoride, hydrocarbons, nickel, electrode carbon, and volatile organic compounds including isocyanates.[42] Fluorine, solids and hydrocarbon emissions to water, and sludges containing carbon and fluoride to land are associated with aluminium production.[5] Metal smelting industries are second only to the chemicals industry in terms of total emissions of toxics to the environment.[5]

Aluminium plants in the UK have been frequently criticised for high levels of discharge of toxic heavy metals to sewers.[9]

Emissions of dioxins have also been associated with secondary aluminium smelting.[33]

Bauxite refining yields large volumes of mud containing trace amounts of hazardous materials, including 0.02kg spent 'potliner' (a hazardous waste) for every 1kg aluminium produced.[43]

Acid Rain

Emissions of sulphur dioxide and nitrous oxides are associated with aluminium production.[5]

Photo-chemical Oxidants

Nitrous oxide emissions are associated with aluminium production.[27]

Other

The open-cast mining of the ore, bauxite, and of the limestone needed for processing can have significant local impact, bauxite mining leaving behind particularly massive spoil heaps.[16]

The association between aluminium smelting and large scale hydro-electric dams in third world countries is well known. So too is the damage such schemes cause to both human communities and to the natural environment.[8,10]

Use

Recycling/Reuse/Disposal

Aluminium is normally easily recycled, saving vast amounts of energy compared to making new, but powder coated aluminium is not recyclable.[3] Powder coating is usually necessary because aluminium's natural corrosion resistance cannot cope with the acidity of most rainwater in the UK.

(b) Cast Iron

Production

Iron and steel production has been ranked as the second most polluting industry in the UK, second only to coke production (of which it is a major customer).[22]

Energy Use

Iron and steel manufacturing requires large amounts of energy, and is one of industry's largest energy consumers.[3]

(See Embodied Energy, p.171.)

99.7% of principle feedstocks used in our iron and steel industry is imported, mainly from Australia, the Americas and South Africa,[18,25,45] and the transport energy costs should be taken into account.

Resource Depletion (bio)

Brazil exports huge quantities of iron to the West, much of it produced with charcoal made from rainforest timber.[6,8] Clearance of land for iron ore extraction in Brazil may also contribute to rainforest destruction.[42]

Resource Depletion (non-bio)

Iron-ore, limestone and coke (made from coal) are required for iron and steel production. Easily available sources of iron ore are getting scarce, and the raw materials have to be transported increasing distances.[3] Proven reserves of iron ore are estimated to be sufficient for 100 years supply if demand continues to rise exponentially, and 200 years at current levels of demand.[46]

Global Warming

The chemical reaction of smelting iron combines the carbon in coke or charcoal with the oxygen in iron(III)oxide to produce CO_2.[22] Due to the size of the industry, global figures for CO_2 emissions from iron & steel production are significant, although much smaller than those from burning fossil fuels (about 1.5%). Nitrous oxides are also produced,[27] and CO_2 emissions incurred during global transport of raw materials (See 'energy use') should also be considered.[42]

Toxics

Recent estimates from the Department of the Environment rate sintering (an early stage in iron and steel production) as possibly one of the largest sources of dioxin emissions in the UK, but there are no reliable figures as yet.[33] Emissions of carbon monoxide, hydrogen sulphide and acid mists are also associated with iron and steel production, along with various other acids, sulphides, fluorides, sulphates, ammonia, cyanides, phenols, heavy metals, metal fume and scrubber effluents.[5,27]

Metal smelting industries are second only to the chemicals industry in terms of total emissions of toxics to the environment.[5] In the UK, iron and steel production tops the chart of fines for water pollution offences.[11]

Acid Rain

Emissions of sulphur dioxide and nitrogen oxides are associated with iron and steel production.[5,27]

Photochemical Oxidants

Emissions of hydrocarbons and nitrogen oxides are associated with iron and steel production.[5,27]

Other

The extraction of iron ore, limestone and coal all has an impact locally.[3]

Use

Durability/Maintenance

Cast iron gutters need regular painting to maintain appearance, but cast iron is in fact reasonably corrosion resistant.[4]

(c) Steel

Steel is produced from further refining basic iron, so for this analysis we have treated it the same as for cast iron with the following additions:

Production

Energy Use

The production of steel from pig iron requires further melting and processing. The embodied energy of steel is 25-33MJ kg^{-1}.[42] (19,200 Btu/lb.)[43]

Resource use (non-bio)

Steel is manufactured using about 20% recycled content, 14% of which is post-consumer.[47]

Global Warming/Acid Rain

Combustion emissions from ore refinement, blast furnace and oxygen furnace operations include greenhouse- and acid rain forming gases. About 3 tonnes of CO_2 are emitted per tonne of steel produced from ore, and 1.6 tonnes per tonne of recycled steel.[42]

Toxics

The refining of steel from iron is associated with further emissions of carbon monoxide, dust, metal fume, fluoride and heavy metals.[27]

Before the recent estimates for dioxins from the sintering process mentioned under Iron above, steel smelting was also listed as a source of dioxin.[14] It would appear that this second source of dioxin emissions arose from the recycling of scrap steel, which is widely used, as a result of PVC and other chlorinated plastic coatings.[1]

Use

Durability/Maintenance

Steel rainwater goods will either require galvanising, other coatings during manufacture, or regular painting to prevent corrosion.

Recycling/Reuse/Disposal

The ease of reclamation is the main environmental advantage of steel.[47] Steel is easily removed from the waste stream magnetically and can be recycled into high quality steel products,[43] and the estimated recovery rate is currently 60-70%.[42] Recycled steel consumes around 30% of the energy of primary production (8-10MJ kg^{-1}

Proof is in the Drinking

The 'Integrated Solar Dwelling', an experimental green building built in the late 1970s at Brighton Polytechnic by John Shore and Francis Pulling met all its fresh water needs from rainwater, including cooking and hot drinks. Using only a settling tank and pumping water when needed to a cold tank at roof level, they had no trouble with water quality, and without even using any filtration.[20]

recycled), including the energy required to gather the scrap for recycling.[42] It is thought that through increasing the extent of recycling and using renewable energy, there is scope for steel to be produced sustainably.[42]

ALERT

Dioxins, released during the manufacture and recycling of steel, have been identified as hormone disrupters.[41] - See PVC 'Alert' section for further detail.

(d) Glass-Reinforced Polyester

GRP rainwater products are made from layers of glass fibres bound by polyester resins. The glass fibres, accounting for about 30% of the product by weight, are made from sand, soda ash and limestone, melted at high temperature. Polyester resins are products of the petrochemicals industry, as are the other additives which are used in smaller quantities.

Production

Energy Use

Both glass-making and petrochemicals are high energy processes. (See Embodied Energy above.)

Resource Depletion (non-bio)

Oil is the main raw material for polyester.[12] Proved oil reserves world-wide will last less than 40 years at current consumption.[19]

Toxics

Emissions of particulates, oils, phenols, heavy metals and scrubber effluents are all associated with petrochemical manufacture. Petrochemical industries are responsible for over half of all emissions of toxics to the environment.[5] Though associated with toxic emissions, the effects from the manufacture of polyester resins are relatively small compared to PVC.[23]

The manufacture of the glass fibres is associated with emissions of a number of toxics, including fluorides, chlorides, fibre particles and volatile organic compounds, and solvent releases.[26]

Dust arising from glass fibre processes can cause skin, throat and chest complaints. Particles of glass fibre may well be emitted to the air from glass fibre production.[26] Manufacturers are recommended to take the same safety precautions with glass fibres as with asbestos fibre[15] and in the USA glass fibres (as used for insulation) narrowly

escaped being listed by the government as carcinogenic due to corporate lobbying.[17] However it appears that the main health risks are associated with insulation fibres, which come in much smaller sizes, than with the continuous filament fibres used for GRP.[28]

Acid Rain

Petrochemical refineries, the source of many of the raw materials for GRP, are major polluters with the acid rain forming gases SO_2 and NOx.[5] Glass fibre production also contributes to acid rain pollution, mainly through the burning of fossil fuels to melt the ingredients.[26]

Photochemical Oxidants

Likewise, petrochemical refineries are responsible for significant emissions of photochemical oxidants such as hydrocarbons,[5] and glass fibre production also contributes.[26]

Other

The extraction of all the raw materials for GRP - sand (not sea-sand), limestone, crude oil - can cause significant local impact.

Use

Durability/Maintenance

The resin of the outer layer of GRP, the 'gel coat', is mixed with pigment during manufacture, and so the product does not normally require painting during its lifetime.

Recycling/Reuse/Disposal

The compound nature of GRP means that it cannot effectively be recycled. Polyester resins are thermosetting plastics, which mean they cannot be remelted by the application of heat.

(d) uPVC

Production

Energy Use

Like most plastics, much of the raw material for PVC is derived from fossil fuels. (Despite this, PVC has one of the lowest embodied energies of plastic materials, at between 53MJ kg$^{-1(34)}$ and 68MJ kg^{-1}.$^{(35)}$) The production of both ethylene and chlorine is energy intensive. Chlorine production in Germany accounts for 25% of all chemical industry's energy consumption (2% of national total).[12] Incineration (not recommended because of toxic emissions) only recovers 10% of this embodied energy.[12]

"An Australian, accustomed to a great variety of surface storage, is astounded that there are no significant domestic rainwater tanks in Europe, the USA, or India (where clean drinking water is rare), ... and that expensive pipelines and bores are the preferred 'alternative', even where local rainfall often exceeds local needs...

"It makes far more sense to legislate for [storage] tanks on every roof than to bring exotic water for miles to towns; it will also ensure that clean air regulations are better observed locally, that every house has a strategic water reserve, and that householders are conservative in their use of water."
Bill Mollison (21)

Resource Depletion (non-bio)

Oil and rock salt are the main raw materials for PVC manufacture[36], both of which are non-renewable resources. One tonne of PVC requires 8 tonnes of crude oil in its manufacture (less than most other polymers because 57% of the weight of PVC consists of chlorine derived from salt).[10]

Proven global oil reserves will last us less than 40 years at current consumption.[19]

Toxics

High levels of dioxins have been found in the environment around PVC production plants, and a Swedish EPA study in May 1994 actually found measurable quantities of dioxins, furans and PCBs in PVC itself.[13] A study of waste sludge from PVC production going in to landfill also found significant levels of these persistent organochlorines.[13]

---

land, but PVC production is top of the list for toxic emissions to all three.[23]

PVC powder provided by the chemical manufacturers is a potential health hazard and is reported to be a cause of pneumoconiosis.[38] High levels of dioxins have been found around PVC manufacturing plants,[13] and waste sludge from PVC manufacture going to landfill has been found to contain significant levels of dioxin and other highly toxic compounds.[12] It was recently reported that 15% of all the Cadmium in municipal solid waste incinerator ash comes from PVC products.[36]

Emissions to water include sodium hypochlorite and mercury, emissions to air include chlorine and mercury. Mercury cells are to be phased out in Europe by 2010 due to concerns over the toxicity,[36] and in 1992 only 14% of US chlorine production used mercury.[36] We found no data regarding mercury in UK production.

Soakaways

"Soakaways have been the traditional way to dispose of stormwater from buildings and paved areas remote from a public sewer or watercourse. In recent years, soakaways have been used within urban, fully-sewered areas to limit the impact on discharge of new upstream building works and to avoid costs of sewer up-grading outside a development. Soakaways are seen increasingly as a more widely applicable option alongside other means of stormwater control and disposal." (Soakaway Design, BRE Digest No 365, 1991)

PVC also contains a wide range of additives including fungicides, pigments, plasticisers (See ALERT below) and heavy metals, which add to the toxic waste production.[15,19] Over 500,000kg of the plasticiser di-2-ethylhexyl phthalate (commonly referred to as DOP, or pthalate), a suspected carcinogen and mutagen (see 'ALERT' below) were released into the air in 1991 in the USA alone.[36] Even 'unplasticised' PVC (the 'u' in uPVC) may contain up to 12% plasticiser by weight.[12]

PVC for outdoor use is usually stabilised with the heavy metal cadmium.[12]

The recent DoE estimates of dioxin sources (mentioned under Iron above) do not mention PVC production specifically.

PVC is manufactured from the vinyl chloride monomer and ethylene dichloride, both of which are known carcinogens and powerful irritants.[12,37] A 1988 study at Michigan State University found a correlation between birth defects of the central nervous system and exposure to ambient levels of vinyl chloride in communities adjacent to PVC factories.[36] In 1971 a rare cancer of the liver was traced to vinyl chloride exposure amongst PVC workers, leading to the establishment of strict workplace exposure limits.[36]

Despite high standards in emissions monitoring and control, large amounts of these chemicals end up released into the environment. When accidents and spills occur, severe contamination results.[12]

The most common situation is when the polymerisation process has to be terminated quickly due to operator error or power faliure, when sometimes the only way to save a reactor from overheating and blowing up is to blow out a whole batch of vinyl chloride.[36]

Her Majesty's Inspectorate of Pollution guidelines indicate that all plastics making processes cause emissions of their raw materials and waste by-products to air, water and

Acid Rain

Petrochemical refineries, the source of may of the raw materials for PVC, are major polluters with the acid rain forming gases SO_2 and NOx.[5]

Photochemical Oxidants

Likewise, petrochemical refineries are responsible for significant emissions of photochemical oxidants such as hydrocarbons.[5]

Other

The extraction and transportation of crude oil can cause significant local impact.

Use

Recycling/Reuse/Disposal

Because of the many different additives present in PVC, it is "impossible to recycle in the true sense of the word".[12] The recycling opportunities for PVC are limited as recycled PVC can only be used for low grade products such as park benches and fence posts. However, post consumer recycling of PVC is currently negligible and some companies actually lose money on every pound of PVC they take.[12,36] PVC also complicates the recycling of other plastics, particularly PET, as it is hard to distinguish between the two. PVC melts at a much lower temperature to PET, and starts to burn when the PET starts melting, creating black flecks in the otherwise clear PET making it unsuitable for many applications. The hydrogen chloride released can also eat the chrome plating off the machinery, causing expensive damage.[7]

Incineration of PVC releases toxins such as dioxins, furans and hydrogen chloride, and only makes available 10% of PVC's embodied energy. 90% of the original mass is left in the form of waste salts, which must be disposed of to landfill.[10] Hydrochloric acid released during incineration damages the metal and masonry surfaces of incinerators, necessitating increased maintenance and replacement of parts.[7] The possibility of leaching plasticisers and heavy metal stabilisers means that landfilling is also a less than safe option.[16]

Health Hazards

All new plastic products have that characteristic 'plasticy' smell - this is caused by off-gassing of the many different constituent chemicals. PVC may release the highly carcinogenic vinyl chloride monomer for some time after manufacture.[2] With PVC rainwater goods it should not be a problem to store and work with them outside to avoid this risk. Waste offcuts or old PVC should never be burnt on bonfires - phosgene, dioxins and hydrogen chloride fumes given off are all extremely dangerous.[2]

PVC can present a serious health hazard during fires, as illustrated by the Dusseldorf airport fire, where welding sparks ignited PVC coated materials and the fire released caustic hydrochloric acid and highly toxic dioxins as well as carbon monoxide and other fumes. The burning PVC also emitted a large amount of dense black smoke which made it difficult for people to escape.[40] Despite Vinyl Institute claims that not one death in the US has been linked to PVC, the US Consumers Union lists several autopsies specifically identifying PVC combustion as the cause of death.[36] Ash from fires in PVC warehouses contains dioxins at levels up to several hundred parts per billion, making a significant contribution to environmental contamination.[36]

ALERT

See Toxics above and also the discussion on page 1.

Pthalates used as plasticisers in PVC, together with dioxins produced during the manufacture and incineration of PVC have been identified as hormone disrupters, and there is convincing, but not definitive evidence linking them to a reduction in the human sperm count, disruption of animal reproductive cycles[39] and increased breast cancer rates in women.[36] Hormone disrupters operate by blocking or mimicking the action of certain hormones. Humans are most affected through the food chain, unborn children absorb the toxins through the placenta, and babies through their mothers milk.[41]

The environmental group Greenpeace is campaigning worldwide for an end to all industrial chlorine chemistry including PVC due to its toxic effects.

British Plastics Federation Statement on the Green Building Handbook's Review of Rainwater Goods

The British Plastics Federation does not accept many of the views and opinions expressed in the Review as they relate to the products for which it is responsible, namely rainwater goods made from Polyvinyl-chloride (PVC) and Glass Reinforced Polyester (GRP). Whilst recognising the need for environmental impacts to be considered we fundamentally question the value of this when it is detached from considerations of the performance of a product in use.

The relevent key facts about PVC are as follows:

It is an old established and well researched material (first synthesised in 1872 and coming into commercial use in the 1920s).

It is widely used in a variety of building applications: rainwater goods; underground pipework; windows; doors; cladding; flooring; wall coverings; and electrical wiring.

PVC has met with wide customer acceptance so much that the European market for PVC construction products grew by 10% in 1994.

At least five eco-balance studies relating to PVC building products have been carried out since 1990 and these demonstrate that PVC holds a favourable ecological position relative to other products.

PVC saves energy because it is 57% based on salt, a naturally occuring substance, and because it is light in weight it consumes little energy during transportation.

Companies manufacturing PVC as a raw material or as rainwater goods operate in well regulated situations and often meet relevant safety legislation and norms by a wide margin of safety.

PVC is recyclable and is being recycled. It can be safely incinerated in modern incinerators. PVC is inert in landfill sites and actually contributes to their stability.

It is true that in some parts of Europe from the late 1980s, some local authorities responded to the alarmist concerns about PVC voiced by some, but not all, environmental groups. This led to, for exmple in Germany, a small proportion of local authorities restricting the use of PVC in publicly subsidised housing. Those measures not only significantly altered the cost structure of public building projects, but over time scientific evidence suggested that they would not yield a particular environmental benefit. Consequently in those instances PVC is gradually being reapproved as an appropriate building material. The manufacture and use of PVC continues to be supported by governments worldwide.

In relation to Glass-Reinforced Polyester (GRP) it is important to recognise that:

Manufacturers of the raw material and products operate in well regulated conditions.

GRP can be recycled and examples of new products are ventilators and electricity meter cupboards. As Plastic Blast Media, recycled GRP can be used as a replacement for chemical paint stripping. Further it can be pyrolised to produce fresh feedstock to the chemical industry in place of oil.

GRP can be safely incinerated and its inherent energy content recovered.

For further information please contact:
Mr Philip K Law
British Plastics Federation
6 Bath Place
Rivington Street
London EC2A 6JE
Tel. 0171 457 5000
Fax. 0171 457 5045

13.5 Alternatives

13.5.1 Timber Gutters

Some historical buildings in the UK use timber gutters and down-pipes. Anecdotal evidence suggests that they are still manufactured and used in parts of Yorkshire. In theory they can be sealed with tars or resins. Given the apparently high environmental impacts of modern rainwater products, widespread moves towards this more traditional material may offer the best options for environmental building in the UK.

13.5.2 Rainwater Storage

If you want to store rainwater for indoor use, it is often a good idea to reject the first few gallons collected from a roof, as it is likely to contain most of the dust and debris settling on the roof in the dry spells inbetween showers. Illustrations in "Permaculture: A Designers' Manual" (below) show two ways how this might be done.[21]

13.5.3 Gargoyles!

For a long time spouts or gargoyles were the standard method of directing rainwater collected on a roof away from a building. They are certainly more fun than the average black drainpipe. Thatched houses also function without rainwater goods and simply project the flow of water away from the building.

13.5.4 Bamboo

Large sections of bamboo, split lengthwise to form gutters, or whole as downpipes, are used in some countries. Preservatives are needed to keep the bamboo from rotting or insect/fungal attack.[7]

13.5.5 Chains

Swiss buildings often used chains instead of pipes to direct smaller water flows down from a gutter spout (presumably because a chain won't get blocked by freezing?). Larger flows would be projected by the spout clear of the building.

13.5.6 Water Storage Roofing

The whole system of gutters, drain-pipes and soakaways can be done away with if you are prepared to consider the idea of using flat roofs for water storage. The collected water is used for flushing W.C.s, and as more water will be needed than the skies provide, no other disposal route will be necessary. Whilst the idea of a flooded roof may seem fraught with problems, its proponents claim that the constant covering of water actually helps to protect waterproof membranes so maintenance may well be lower than with ordinary flat roof designs, and the weight of the water is spread evenly over the whole structure, so loading is not a particular concern. The Green Building Digest magazine (issue 14) and AECB's Greener Building directory goes into the idea in more detail.[8]

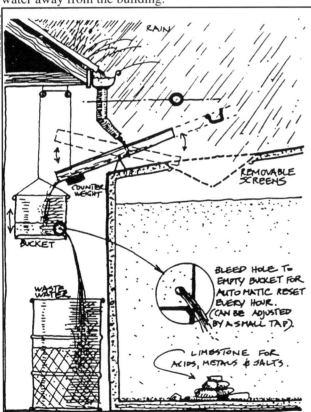

METHODS OF REJECTION OF FIRST WATER FLOW FROM A ROOF.
The first rains wash the roof and are rejected; these systems automatically reset when empty.

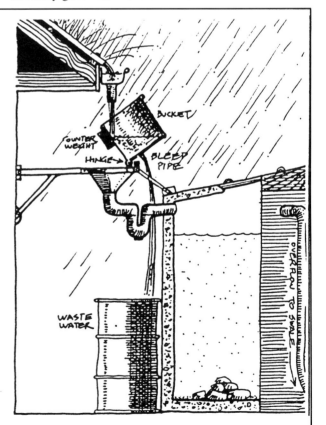

Illustration reproduced with permission from : Permaculture: A Designer's Manual, Bill Mollison 1988. Tangent Publications

13.6 References

1. Dioxins in the Environment, Pollution Paper 27 (Department of the Environment, HMSO, London, 1989)

2. C for Chemicals (M Birkin & B Price, Green Print, London, 1989)

3. The Green Construction Handbook - A Manual for Clients and Construction Professionals (J T Design Build, Bristol 1993)

4. Construction Materials Reference Book (ed. D K Doran) Butterworth-Heinemann Ltd, Oxford, 1992

5. The World Environment 1972-1992 - Two Decades of Challenge (eds. M.K. Tolba et al) UNEP/Chapman & Hall Ltd, London, 1992

6. Eco-Renovation - the ecological home-improvement guide (E Harland, Green Books Ltd, Dartington) 1993

7. Appropriate Building Materials - A Catalogue of Potential Solutions (Third Revised Edition) (R Stulz, K Mukerji, SKAT Publications, Switzerland) 1993

8. Greener Building Products & Services Directory (K Hall & P Warm) Association for Environment Conscious Building, Coaley

9. The Secret Polluters (Friends of the Earth, UK, July 1992)

10. Environmental News Digest vol.10 no.1 (Jan. '92, Friends of the Earth Malaysia)

11. ENDS Report March 1994 Issue No. 230 Page No. 44 (Environmental Data Services)

12. PVC - Toxic Waste in Disguise (Greenpeace International, Amsterdam) 1992

13. Achieving Zero Dioxin - An Emergency Strategy for Dioxin Elimination (Greenpeace International, London) 1994

14. Chlorine-Free Vol.3 No.1 (Greenpeace International 1994)

15. Green Design (Fox & Murrell, Architecture Design and Technology Press, London 1989)

16. Ecological Building Factpack (peacock & Gaylard, Tangent Design Books, Leicester 1992)

17. Rachel's Hazardous Waste News #367, December 9, 1993 (Environmental Research Foundation, Annapolis, MD USA)

18. Minerals Handbook 1992-93 (Philip Crowson, Stockton Press, New York 1992)

19. Oil & Energy Trends Annual Statistical Review (Energy Economics Research, Reading, 1990)

20. Building for a Future, Autumn 1994 Vol.4 No.3 (Association for Environment Conscious Building, Coaley)

21. Permaculture: A Designers' Manual (Bill Mollison, Tagari Publications, Tyalgum, Australia 1988)

22. Environmental Chemistry 2nd Edition (Peter O'Neill, Chapman & Hall, London, 1993)

23. Production & Polymerisation of Organic Monomers IPR 4/6 (Her Majesty's Inspectorate of Pollution, HMSO, London, 1993)

24. Review of Accidents Involving Chlorine (G. Drogaris, Community Documentation Centre on Industrial Risk, Commission of the European Communities, Luxembourg, 1992)

25. UK Minerals Yearbook 1989 (British Geological Survey, Keyworth; Nottingham 1990)

26. Glass Fibres & Non-asbestos Mineral Fibres IPR 3/4 (Her Majesty's Inspectorate of Pollution, HMSO, London, 1992)

27. Metal Industry Sector IPR 2 (Her Majesty's Inspectorate of Pollution, HMSO, London, 1991)

28. Buildings & Health - The Rosehaugh Guide to the Design, Construction, Use & Management of Buildings (Curwell, March & Venables, RIBA Publications, 1990)

29. Environmental Building B4 (Pat Borer, Centre for Alternative Technology, Machynlleth, 1994)

30. Energy Efficient Building: A Design Guide (Susan Roaf and Mary Hancock, Blackwell Scientific, Oxford, 1992)

31. Chartered Institute of Building Service Engineers - Building Energy Code 1982

32. Dictionary of Environmental Science & Technology (Andrew Porteus, John Wiley & Sons, Chichester, 1992)

33. ENDS Report January 1995 Issue No. 240 (Environmental Data Services)

34. Dictionary of Environmental Science and Technology (A. Porteus). John Wiley & Sons. 1992

35. The Consumers Good Chemical Guide. (J. Emsley). W.H. Freeman & Co. London. 1994

36. Environmental Building News Vol.3 No.1. Jan/Feb 1994

37. H is for EcoHome. (A. Kruger). Gaia Books Ltd, London. 1991.

38. Greenpeace Germany Recycling Report. 1992

39. Greenpeace Business no.30 p5, April/May 1996

40. Building for a Future Vol.6 No.2. Summer 1996.

41. Taking Back our Stolen Future - Hormone Disruption and PVC.

42. Environmental Impacts of Building and Construction Materials. Volume C: Metals. (N. Howard). Construction Industry Research and Information Association , June 1995.

43. The Greening of the Whitehouse. http://solstice.crest.org/environment/gotwh/general/materials.html

44. Environmental Protection Act 1990 Part 1. Processes prescribed for air pollution control by Local Authorities, Secretary of State's Guidance - Aluminium and aluminium alloy processes. PG2/6 (91). (Department of the Environment) HMSO 1991.

45. UK Iron and Steel Industry: Annual Statistics. (UK Iron & Steel Statistics Bureau). 1991

46. Pollution: Causes, Effects and Control. 2nd Edition. (R.M. Harrison). Royal Society of Chemistry. 1990.

47. Environmental Building News Vol.4 No.4. July/August 1995.

Toilets and Sewage Disposal   14

14.1 Scope of this Chapter

This chapter looks at the main developments in thinking regarding the more ecological disposal of household sewage wastes. It does not intend to act as a buyers guide to specific makes and models of toilet, nor as a practical guide to the installation or construction of particular models. Such guides already exist and this report aims both to point to these other guides where appropriate and to provide an overview of the key issues in the field.

14.2 Introduction

From the householder's point of view, there is nothing more convenient for the disposal of human wastes than the water closet. This is evidenced by the fact that almost everyone has one - with 96% of UK population currently served by mains sewers and the rest connected to septic tanks.[14] The WC is hygenic, odourless, reliable, simple to operate and requires very little maintenance. To some extent the presence of sewers and flush toilets has been used as an measure of a society's 'progress' and affluence. Nevertheless, with increasing value being given to ideas of 'sustainability', it appears that in the UK the disposal of wastes once they have passed beyond the u-bend is far from satisfactory.

14.2.1 Sustainable Sewage?

Before we identify exactly what the problem is, we will briefly outline what the goals of a sustainable sewage system might be.

1. The complete recycling of valuable nutrients contained within human wastes.
2. Careful use of resources (especially water) at a rate which does not exceed their replacement by natural processes.
3. The safe neutralisation of pathogens in human wastes in a manner which does not cause the release of other toxins into the environment.

14.2.2 The Current Position

(a) Recycling of Nutrients

Sewage sludge accounts for around 20-30 million tonnes annually, or 8% of total UK waste arisings.

Disposal routes:[32]

Farmland	46%
Sea dumping	28%
Landfill	13%
Incineration	7%
Other	6%

On the face of it, the figure of 46% 'disposed' to farmland would appear to show that at least some of the nutrients are being recycled. However the root of the UK's sewage problem lies in the fact that human wastes are mixed with all sorts of industrial effluents before they arrive at the treatment plants. Industrial releases, animal and vegetable processing, and storm-water run-off from roads and paved areas all contribute to the potential toxic or pathogenic load of sewage. Toxic heavy metals are of most concern as they will survive all the usual treatments such as pasteurising, composting or filtering.

The use of sewage sludge on agricultural land therefore, whilst of immense potential benefit, is closely regulated to ensure that the build up of pollutants does not pose unacceptable risks.[17] Both the sludge and the soil must be regularly sampled and analysed, and there are limits on applying sludge to crops such as soft fruit and vegetables. (Dedicated sewage 'farms' may apply sludge more liberally, but may not produce food for human consumption.) There are also possible problems associated with transport by tanker, odour control, surface run-off and water pollution.[17]

On the basis of the above evidence, it would appear that considerably more than 56% of nutrients from sewage sludge is wasted, and that our present system is only managing to compensate for this waste by the extensive use of petroleum-based fertilisers.[19] The use of composted human waste on crops grown under organic schemes is currently prohibited due to concerns about heavy metals.

(b) Water Use

"It is highly likely that water, like energy in the 1970s, will become the most critical resource issue in most parts of the world by the late 1990s and the early part of the twenty-first century"
(M K Tolba[4])

UK Domestic Water Consumption (Source 30)

Washing Machine	**12%**
Bath/Shower	**17%**
Outside	**3%**
Toilet Flushing	**32%**
Other	**32%**

There are two key issues regarding water use for green commentators which are dealt with later in this report. Toilet flushing is the largest single domestic use of water and in the event of shortages, toilets should be able to contribute to reducing their impact. Second is the fact that water used to flush toilets is 'drinking quality' which may indicate a waste of resources in treating water to a higher standard than necessary.

Domestic water supplies are usually unmetered, and charged for on the basis of rateable value. This method was to be ended by the year 2000 in order to restrain water use, but this date has now been extended indefinitely.[14] Future charging options are likely to include more widescale metering, or charges based on Council Tax or other types of property banding.[14]

(c) Safe Neutralisation of Wastes

The modern mains supply of clean (drinking quality) water is a complex process. For example, in London water is extracted by pumping station from below a dam in the river Thames, pumped to a reservoir for storage and settlement, sent through an aeration basin, primary filters, through sand or other filters, and then sterilised with chlorine injection before further pumping, storage and distribution through the water mains. Throughout this process the water quality is monitored in terms of the E. Coli count, an indicator of faecal pollution.[2]

This system is not perfect, and flooding from sewers, although infrequent, happens to around 0.1% of properties. Sewers also overflow during storms and tests of drinking water quality still turn up coliforms/faecal coliforms - indicators of sewage contamination.[14]

Moreover, whilst EC Directives require all significant sewage discharges to be treated, and prohibit the dumping of sewage sludge at sea by 1998, perhaps 50% of discharges into tidal waters remain untreated at the time of writing.[14]

More details of land, air and water pollution from UK sewage systems appear on pages 4 and 5, which also mentions some criticisms of the use of chlorine as a purifier which we have discovered.

14.2.3 Conclusions

"The persistence of metals such as zinc, copper, cadmium, nickel, and chromium in soil is a real problem in the present disposal of sewage sludge on land; for example, lead would stay in the ground indefinitely. It is essential in the interests of public health that industry stop discharging waste contaminated with heavy metals into the general sewerage system."
(British Medical Association[5])

It is clear that the UK's current sewerage system cannot be said to be sustainable in any proper sense of the word. Nevertheless a number of green practitioners and designers have been working on a range of solutions which are summarised on the 'at a glance table' opposite. More specific details about the impacts of each system are analysed in page 185 and more detailed descriptions of each system appear on subsequent pages.

However, even in a green future, urban homes are likely to require mains sewerage. There is a strong case nevertheless for an upgrading of the mains sewerage systems on ecological lines. The most important change would be to separate the sewerage of domestic (human) wastes from the disposal of industrial effluents and storm water run-off from roads etc. - 'pure' sewage being much easier to treat and return to the land as fertiliser.

To some extent this is a social or political solution that may be out of the hands of most readers of the Green Building Handbook, except in as much as they are voting citizens and consumers of the products of polluting industry. Nevertheless there is still much that can be done at a local/planning level to look at local ecological treatment systems of the type outlined in 'alternatives'

Finally, the fact that systems of sewage 'treatment' need not be unsustainable is also evidenced by experiences overseas. In Sweden, for example, WC's have actually been banned for new developments in some areas, and biological treatment methods have long been pioneered in the USA.[27]

Some of the most instructive systems in terms of sustainability, however, occur in 'less developed countries'.

"Around the large cities in Japan, South Korea and China there are vegetable-growing greenbelts that rely on [human] wastes. They are either applied directly to the soil after having been left to mature, or treated sewage water is pumped directly onto the fields. Shanghai actually produces an export surplus of vegetables. And in India 'nightsoil' is the basic component of Calcutta's aquaculture system, producing 20,000kg of high protein fish for sale every day".[31]

"What people do with their shit probably relates more immediately to people's psychological approach to recycling than any other issue to do with buildings."
(Heimir Salt - Green Building Digest Editorial Panel!)

Urban Housing

Dry composting toilets generally require land on which to spread the compost and to disperse liquid effluents. This makes them less suitable options for most urban situations.

Low Flush WCs

(see Greenie Points p182 and Product Details p189)

Grey Water Recycling

Grey water is the water from baths ,sinks, showers etc. It can be 'recycled' for flushing toilets.
(see Alternaives p191)

Community Scale Developments

Solar aquatics and other more complex ecological treatment methods may not be cost effective for single households - but are ideal for larger developments.

Local Ecological Water Treatment

Biological sewage treatment systems using methods like reed-beds and solar aquatics are now proven to be just as effective as standard chemical treatments.
(see Alternatives p190)

Twin Vault

Cheap self-build variation on the earth closet pioneered in the UK by the Centre for Alternative Technology (CAT) and based on a Vietnamese design.
(see Product Details p.187/189)

Rural/Remote Locations

Dry composting toilets make an ideal alternative for remote situations where the only options are septic tanks or cesspools.

'Dry' Composting Toilets

C.A.T argues that because composting toilets are still a relatively new technology, the more cautious may prefer to run them alongside an exisiting septic tank set-up before disconnecting from mains sewerage entirely.[1] There may also be legal considerations.
(see Practicalities p192)

De-watering Type

Collects and dries wastes for composting elsewhere. A more compact variety but can require more frequent emptying and a higher electricity consumption. There are also **urine separating versions** *which collect and store urine separately for later use as a fertiliser.*
(see Product Details pp187/189)

Multrum Type

More sophisticated (and expensive) variation on the earth closet.
(see Product Details pp187/189)

Eco-Sewage - At a Glance Guide

	£	Toilets & Sewage Disposal									
	Unit Price Multiplier	Unit Size	Frequency of User	Maintenance etc	Water Use	Electric Power Use	Pollution of Land	Pollution of Water	Pollution of Air	Toxics in Treatment Process	Sustainability/Nutrient Loss
Wet Systems											
WC + sewer & public treatment plant	1	•			●		●	●	●	●	●
WC + septic tank & land drain	1.4	•			●		●	•	●	•	●
WC + local ecological treatment (sludge composting, reed beds, solar aquatics etc)	0.7	•			●						
Dry Systems											
Composting toilet - Twin-Vault 'C.A.T.' type	0.6	●		•							
Composting toilet - Commercial 'Multrums' etc.	2.8	●		•		•					
Composting toilet - De-watering type	1.5	•	●			●					

Greenie Points

The new Environmental Standard version of BREEAM for new homes[13] awards one optional point for specifying that all WCs have a maximum flushing capacity of 6 litres or less, but notes that as low as 3.5 litres is possible. (Water Byelaws' maximum is 7.5 litres).

"*We assume that by flushing and forgetting we are rid of the problem, when we have only compounded it by moving it to another place. Every tenderfoot camper knows not to shit upstream from camp, yet present urban culture provides us no alternative..... The waste we seek so hard to ignore threatens to bury us.*"

Sim Van der Ryn (19)

14.3 Product Analysis

The analysis below gives further details about the criticisms noted on the Product Table.

(a) WC + sewer & public treatment plant

Unit Price Multiplier

Where water is metered, the running costs of a WC can be high: with water costing from 41p to 75p per 1,000 litres, and sewerage charges (which assume 90% of water used goes down the drain) between 41p and £1.57 per 1,000 litres (of water used?).[18] To calculate the unit price multiplier we have taken the mean of these charges and assumed an average households usage at 112 litres per day. We have also added a mean purchase price of a standard WC suite (unfitted) at £104.

Water Use

Water use averages 140 litres per person per day, the largest single usage being the more than 30% used to flush toilets.[14]

There can be high levels of abstraction of water from rivers in some areas, and the NRA has identified 40 rivers where reduced flow is a problem.[14]

Pollution of Land

The use of sewage sludge on agricultural land is closely regulated to ensure that the build up of pollutants does not pose unacceptable risks.[17] Both the sludge and the soil must be regularly sampled and analysed, and there are limits on applying sludge to crops such as soft fruit and vegetables. (Dedicated sewage farms may apply sludge more liberally, but may not produce food for human consumption.)

13% of sewage sludge in the UK is currently landfilled, (see above). Presumably this sludge is too toxic for use on agricultural land.

Pollution of Water

Sewage effluent standards for the UK were set in 1915 - but in 1989 up to 20% of sewage treatment plants did not meet these.[2] Sewage treatment works may produce phosphorus inputs to fresh water responsible for blue-green algal blooms.[14]

Sewage sources account for the largest proportion of water pollution incidents by type (28% in 1991).[14] Overall non-compliance with discharge consents is around 12% (with regional variations) but declining.[14]

Pollution of Air

7% of sewage sludge in the UK is currently incinerated. Pollution caused by the incineration of sewage sludges includes: particulates, heavy metals. sulphur, nitrogen and carbon oxides, halogen compounds, dioxins and organic compounds to air; mercury, cadmium and PCBs in effluents to water; halogens, organo-metallic compounds, dioxins, furans and PCBs and other heavy metal compounds in ashes and residues taken to landfill.[15,16]

Sewage incineration can cause air pollution - the following are controlled by legislation: Carbon monoxide, 'organic compounds', particulates, heavy metals, chloride, fluoride and sulphur dioxide.[15]

Toxics in Treatment Process

The chlorination of water is not without health and environmental risks of its own making. It may combine with other chemicals to produce cancer causing agents and chloroform in drinking water; chlorine in effluent dumped at sea may form toxic acids; chloroform may even be released into the atmosphere affecting the ozone layer.[19]

Greenpeace have recently been campaigning for an end to chlorine chemistry internationally.[28,29]

(b) WC + Septic Tank & Land Drain

Unit Price Multiplier

Based on the following costs: septic tank for 4 person household ~ £450 plus sitework. Emptying ~ £60 every 3-5 years. Added to this were the water use (but not sewage) charges, and purchase price based on the calculations above for 'WC + Sewer'.

Water Use

A septic tank will use the same quantities of water for flushing as for a 'mains' WC. (See the comments above for Water Use under WC+sewer etc.)

Pollution of Land/Pollution of Air/Sustainability & Nutrient Loss

We have not been able to discover any discussions of final destinations for sewage sludge removed from septic tanks. We have therefore assumed for the purpose of this report that disposal of septic tank sludge will be distributed in the same proportions (ie. to landfil, agricultural land, incineration etc) as other sewage sludge. For specific criticisms of these impacts see WC + Sewer & Public Treatment Plant above.

(c) WC + Local ecological treatment

Unit Price Multiplier

The only reference to the cost of solar aquacell plants that we have discovered suggested that 'construction and operating costs are half that of secondary sewage treatment'.[19] We have based our calculation on the assumption that these savings would be passed onto individual householders in an annual sewage treatment charge at half the current rate. Water use and WC purchase costs are the same as those above.

Water Use

An ecological treatment process based on WCs will use the same quantities of water for flushing as for a 'mains' WC. (See the comments above for Water Use under WC+sewer etc.)

(d) Composting toilet - Twin-Vault 'CAT' Type

Unit Cost Multiplier
Rough costs have been calculated on the basis of standard building costs for excavation of pits, establishment of a concrete base, standard block construction around the vaults and a timber housing. It may well be possible to construct such toilets much more cheaply since some commentators have suggested that, in terms of overall costs, including construction and maintenance of the whole system, simple latrines (similar in concept to composting toilets) are on average less than one tenth the cost of conventional flush toilets and sewerage.[22]

Unit Size
This system usually requires about two cubic metres below the toilet for storage of the waste, and an area for access/emptying.[1]

Frequency of User Maintenance
The chambers are sized according to the expected amount of use, so that by the time the second chamber is full, the sewage in the first chamber is composted. If the whole cycle takes two years in British conditions, one chamber will need emptying every twelve months.[1]

(e) Clivus Multrum type composting toilets

Unit Price Multiplier
We have added the purchase price of a four person Clivus[27] to the estimated energy costs of running a 34 watt fan continuously.

Unit Size
The main disadvantage of Multrums is their size. A 3-4 cu.m version will occupy space of 3.2m by 1.2m, ie.the full height of a room or cellar, plus an area to access/empty etc.

Frequency of User Maintenance
Clivus Multrums are designed to require emptying less frequently than once a year. (15-24 months).

Electric Power Use
Most Clivus type models now come with a 34 watt electric fan to remove odours and help aeration.[27]

(f) Composting Toilet - de-watering type

Unit Price Multiplier
We have added £609 (the mean purchase price of a four person dewatering toilet)[27] to the estimated energy costs of running a 34 watt fan continuously and the energy costs of running a 300 watt heater for two hour per day. If the heater element requires a higher rate of usage (see below) then the Unit Price multiplier could easily become comparable with those of the more expensive Multrum above.

Unit Size
Although much smaller than Multrums, de-watering toilets are still larger than a standard WC and may have trouble fitting into existing spaces.[1,27]

Frequency of User Maintenance
The various models of de-watering toilet require emptying with a frequency of between 2 and 18 weeks.[27] The average emptying time will be around 8 weeks.

Electric Power Use
All de-watering toilets use an electric fan to remove odours and help aeration.[27] Many also use a heating element of usually around 300 watts, and the frequency with which it will be used depends upon the ambient temperature of the house it is situated in.[27] According to C.A.T. 'in some circumstances the toilet can become the largest consumer of electricity of all the household appliances. From an environmental point of view we have to set the impact of this energy consumption against the other benefits, and the verdict is generally negative'[1]

Handling human wastes can involve health and pollution hazards and should not be conducted without a good understanding of the key issues. These are well covered in the books 'Water Treatment and Sanitation'[33] and 'Small Scale Sanitation'.[22] The information above is only intended as a brief overview and a guide to further materials.
(See also Specialist Suppliers)
Drawings overleaf are reproduced with permission from the Centre for Alternative Technology's book 'Fertile Waste'. Details of how to get hold of it appear in Recommended Further Reading.

14.4 Product Details

14.4.1 Composting Toilets -

Although composting at high temperature kills most pathogens (disease spreading bacteria etc.) most composting toilets will not achieve these temperatures. However time is also a good destroyer of pathogens, and after only 6 months it is reckoned that the contents of a closed receptacle will have no higher faecal coliform levels (the standard measure of faecal infection) than normally found in garden soil.[21]

(a) Twin Vault 'CAT' Type

This system usually comprises toilet seats situated directly above two chambers built of concrete blocks, with external access for turning or removing compost, and a shelter for housing the seats (See Picture) Two chambers are built so that when one is full its contents may be aged/composted in situ, thus avoiding the need to handle untreated wastes. A small amount of dry carbonaceus material is added with each use of the toilet, such as sawdust, leaves or grass clippings.[19]

This sort of design does not attempt aerobic composting, which according to C.A.T 'rarely seems to work anyway', but is cheap and simple to build, reliable, needing little maintenance, no handling of uncomposted waste, no energy inputs and 'makes lovely compost'.[1]

The one possible drawback is drainage: there may be surplus liquids, which need to be drained off. Simply soaking into the ground around the toilet, or collecting in buckets to be used diluted on the garden are possible, but not ideal answers. This liquid may contain pathogens, so care is needed.[1] A better idea is to make sure things are kept dry, by using plenty of 'soak' (sawdust etc), and better still, by excluding urine. It is possible to design a toilet seat (or squatting hole) that without too much bother can separate flows of urine and faeces at source, or it is possible even to install a urinal for male users leading into a suitable container for garden fertilising.[1]

There may be planning problems with this type of installation - see below.

Full details of designs and variations appear in CAT's book 'Fertile Waste'[1] available for £3.95 (+p&p) from Centre for Alternative Technology, Machynlleth, Powys SY20 9AZ.

Twin Vault 'CAT' Type

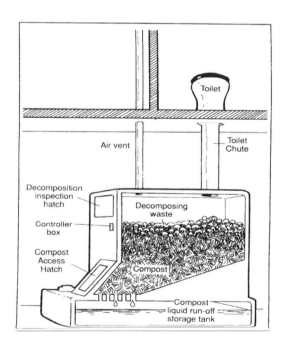

Clivus Multrum Type

(b) Clivus Multrum Type

Originally developed for Swedish summer-houses, remote from mains sewerage and with geology unsuitable for pits or septic tanks. They are among the most tried and tested of composting toilets and comprise a large plastic vessel, with internal baffles shaped to allow gradual sinking of composting material to the bottom in a continuous rather than batch (twin-vault) mode. They also take kitchen wastes and are ventilated through a pipe to the roof, now always fan assisted. Manufacturers claim that because of the lower pressure inside the vessel, there are no odour control problems. The claim is not universally accepted and some have suggested that odours may be a problem on still days -and that Mutlrums are best sited away from the house.[7] Others have described the system as the "method most likely to be socially acceptable in European areas of high density".[6]

Cold may inhibit the digestion process and electric heaters have been used sometimes.[6] Multrums are usually emptied every 15-24 months and come in a variety of sizes for from 4 to 80 people.

The compost from a Multrum is a well balanced, valuable plant fertiliser, its organic structure enabling close on 100% utilisation unlike chemical fertilisers which can be washed away by rain.[21]

More information about current multrum models commercially available in the UK appear in an excellent report in the Spring/Summer 1995 edition of 'Building for a Future ' - available for £3.50 from the Association for Environment-Conscious Building, Windlake House, The Pump Field, Coaley, Gloùcestershire GL11 5DX.

(c) De-watering Type

These types of toilet dehydrate solid wastes, reducing them to around 10% of their original volume. They also evaporate liquids, sometimes via a charcoal filter. They usually use fans driven by small electric motors, and some types have heating elements as well. Electrical consumption is usually around 20-30 Watts for a fan and 300 Watts for a heater. Toilets with heating elements don't usually need drainage for excess water.[27] (Some models run on 12 volts, and so may be suitable for running off a small photovoltaic panel or wind generator.) A family of four would have to empty such a toilet perhaps every 10-12 weeks - usually a simple process, but the residue still needs to be composted in the garden.[7] Some models (e.g. Biolet NE) have two interchangeable containers, which allows the waste to continue drying out and breaking down before being emptied, much like a small scale version of the twin-vault composting toilet.[27]

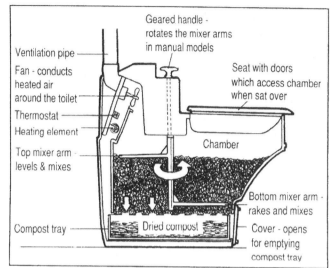

De-watering Type

Ventilation pipe

Fan - conducts heated air around the toilet

Thermostat

Heating element

Top mixer arm levels & mixes

Compost tray

Geared handle - rotates the mixer arms in manual models

Seat with doors which access chamber when sat over

Chamber

Dried compost

Bottom mixer arm - rakes and mixes

Cover - opens for emptying compost tray

Urine Separating Versions

Some models (e.g. Ekologen) collect and store liquids separately. This is a benefit for two main reasons. First, they do not need a heating element as the compost material will be dry. Second, urine is pathogen free and nutrient rich and can function as a useful fertiliser with little further treatment other than dilution. In future urine may well become an essential agricultural nutrient and it is already collected from some apartment blocks in Sweden

More information about current de-watering and urine separting models commercially available in the UK appear the Spring/Summer 1995 edition of 'Building for a Future ' (see above for details).

14.4.2 Low-Flush WCs

Water bye-laws set the maximum size of flush from a toilet cistern as 7.5 litres. (It was higher in the past.) However, there are proven systems capable of operating at only 6 or even 3.5 litres per flush. Dual flush systems have failed to live up to their promise and are currently banned for installation and use with a mains water supply on the basis that some users were not using them properly.[12] Many dual-flush cisterns used more the 9 litres for a full-flush and would now be prohibited.[13]) Other commentators suggest the lower technology solution of putting a brick or a water filled jar/bottle in the cistern to save water. They suggest experimenting to get the right amount).[7]

14.5 Alternatives

14.5.1 Local Ecological Water Treatment(s)

Detailed descriptions of such systems are beyond the scope of this report. More information is available from a comprehensive list of contacts provided by C.A.T. in their 'Sanitation Resource Guide' (Cost 70p + p&p) - address on p192.

(a) Solar Aquatics

"How many people know where water comes from and goes to beyond the limits of tap and plughole? Used water could be cleaned of pathogens, excessively available nutrients and even chemical pollution by a biological and rhythmic flow system of flow-forms, ponds, reed beds and other vegetation. Such systems don't need to be shut away in sewage farms; they can be attractive, even artistic."

Christopher Day, Places of the Soul (8)

This biological system comprises a series of ponds and tanks contained in a large greenhouse to maintain a warm, solar-powered temperature. Aeration and sand filtration are combined with the culture of floating aquatic plants such as water hyacinth, and with a variety of fish, shell-fish and invertebrates. Such systems are better at removing nutrients and toxins than conventional sewerage treatment, and work fine in cooler climates. Costs both for construction and running (and energy) expenses are lower than for conventional sewage. Solar aquatic treatment plants are well suited to smaller scale, community-sized installations and 10,000 people's wastes can be treated with an area of around one acre. There is the added bonus of a crop of fish, shrimp and aquatic plants.[19]

(b) Reed Beds

Constructed reed beds consist of one or more vessels containing gravel and/or sand or soil, with reeds and other aquatic plants growing on top of this. Dirty water flows through the bed, to emerge cleaned at the other end. Large numbers of micro-organisms grow in the root area of the reeds, digesting the slime, nutrients and other pollutants. Reed beds deal with dirty water only, not with faeces in sewage, and so primary treatment, for example in a septic tank or settlement lagoon, is required for this. They might be appropriate for dealing with 'grey water' from houses with dry composting toilets, or for treating the effluent of a septic tank where local conditions do not allow for land drainage. Costs and sizes vary depending on design (there are two types - vertical and horizontal flow - often used in combination). Somewhere between 1 and 2 cubic metres of reed beds are usually needed per person.[23]

"Reed bed systems resemble marshland and are environmentally very acceptable. Over 100 schemes exist in Germany and in the UK - several water companies are running experimental schemes at sewage treatment works."[30]

"The current technology of "waste disposal" (the term reveals the syndrome) is still fighting a war against nature, built on fragments of nineteenth century science not yet integrated into an understanding of life processes as a unified, but cyclical whole."

Sim Van der Ryn (19)

15.5.2 Re-using 'Grey water'

This process aims to save water by installing a tank to store 'grey' water used for washing and bathing and use it for flushing a conventional WC. A simple 'rainmiser' type diverter and a tank of about 30-40 litres capacity are the main ingredients.[7]

Some commentators assert that storing grey water for reuse would require some form of water treatment and that the "minimum acceptable is likely to be filtration and chlorination".[12] This is because organic matter in the stored water may putrefy if it were left for any length of time. Because 'treatment' may involve the use of chemicals, biological treatments with more complex approaches like reed beds (see opposite) may only be suitable for larger developments. Although no regulations cover this, it has been suggested that the Environmental Health Department should be consulted.[12]

Variations can include a larger tank used to supply outside taps for watering and/or a (hand-) pump to take water back up to upstairs WCs.[9]

Another simpler option is a 'basin-topped cistern', where hand washing water ends up directly in the cistern.[9]

The Cambridge Autarkic House[6] used water for toilet flushing, but this water had been collected first as rain from the roof, then filtered and disinfected for use as drinking or washing water, then as 'grey water' collected (with heat recovery) and cleaned for recycling by reverse osmosis - the clean fraction being recycled for washing and the dirty fraction finally stored for use as flushing water.

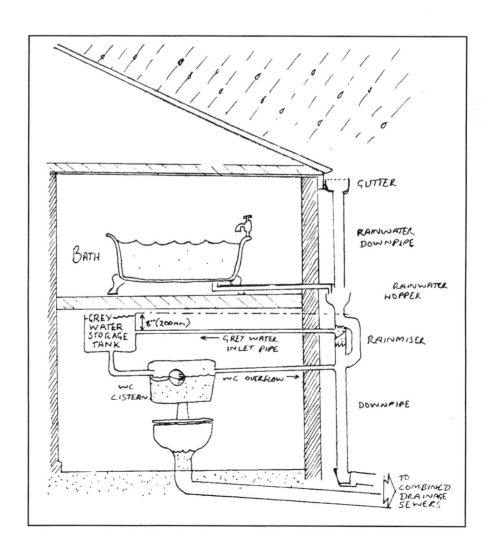

14.6 Practicalities

- **Regulations**
- **Planning law**
- **Water bye-laws**

Regulations require 'adequate sanitary conveniences' in either a bathroom or separate toilet.

Chemical or other means of treatment are permitted where drains and water supply are not available. (Further than 30m/100 feet)

Section 21 of the Building Act stipulates that earth closets, chemical toilets and the like are suitable only as second or temporary toilets. Few building officers have experience of ecological sewage treatment methods and some will be more flexible than others.(27)

Section 66 of the Building Act 1984 enables a local authority to serve a notice requiring replacement of any closet that is not a water-closet - but only where sufficient water supply and sewer are available. (LA must pay half if notice is served.)(10)

Of self-contained eco toilets, only Servator Ecological Toilet (previously known as Lectrolav) has a recognised BBA certification.(27)

Water bye-laws may also affect some models that use a small-volume of water for flushing but don't use a syphonic mechanism, but using rain or grey water from a tank avoids this problem(27).

14.7 Best Buys/ Conclusions

If you want a feeling of hygiene (at least locally), reliability, ease of use, minimum maintenance for the householder, and also want to dispose of grey water off-site, then there is no real competition to the standard WC.(1) To conserve fresh water resources you should fit the lowest water-use model you can find (3.5 litres per flush is possible - see supplier listings) and consider storing rain-water or even bath-water for flushing.

The ecological treatment of sewage from WCs and other grey-water sources, with settlement/septic tanks, sludge composting, and reed beds or solar aquatics is becoming a well established technology. Due to the need for management and maintenance, such systems are perhaps best suited to a community rather than household scale, although they may also be an option for remote households where septic tank land-drainage is not possible.

However, if you really wish to minimise environmental impact then there are practical alternatives such as the composting toilets discussed in this report. The Centre for Alternative Technology(1) recommends staying connected to the main sewer whilst you develop alternative systems. Although experience is growing in this country, many alternatives are still somewhat experimental.

"If there comes an outbreak of typhoid as often as not we find the drains to blame; but as a matter of fact we prescribe more drains as a remedy."
Dr. Vivian Poore M.D.[20]

14.7 Recommended further reading

Fertile Waste - Managing your Domestic Sewage (P Harper & D Thorpe, Centre for Alternative Technology, Machynlleth 1994)
Includes full details of designs and variations.
Available for £3.95 (1.00 p&p) from Centre for Alternative Technology, Machynlleth, Powys SY20 9AZ.

'Building for a Future'
(Spring/Summer 1995 edition, AECB)
Includes specifications and prices for most of the commercially available composting toilets in the UK.
Available for £3.50 from the Association for Environment-Conscious Building, Nant-y-Garreg, Saron, Llandysul, Carmarthenshire SA44 5EJ.

The Toilet Papers - Recycling Waste and Conserving Water (Sim Van der Ryn, Ecological Design Press, California 1995)
Inspirational and polemical deep green analysis.
ISBN: 0-9644718-0-9 (Also available from the Centre for Alternative Technology)

14.8 Suppliers

BIOLET & VERA Composting toilets ECOLOGEN Urine Separating Toilets
EASTWOOD SERVICES, Kitty Mill, Wash Lane, Wenhaston, Near Halesworth Suffolk IP19 9BX　　　(tel: 01502 478 249 fax: 01502 478 165)

UK agent for a number of manufacturers of 'composting and low water use toilets'. Also offers a consultancy service.

Contact: Adam East

COMPOSTER Composting Toilet DE TWALF AMBACHTEN, De Bleken, HB Boxtel, Holland 2,5282 +31 41 167 2621 'Compact Composter' for use in homes, caravans etc. Twin chamber and separating system and DIY kits/plans.

BIOLET Composting toilet WENDAGE POLLUTION CONTROL LTD, M G Mansfield Rangeways Farm, Conford, Liphook Hampshire GU30 7QP 01428 751296 01428 751541 Importers of the 'Biolet' range of composting toilets. Models available to suit domestic or camping facilities.

CLIVUS MULTRUM Composting Toilets
CLIVUS MULTRUM INC, 104 Mt Auburn St, Harvard Sq, Cambridge, Massachusetts 02138, USA (tel: 001 508 725 5591 fax: 001 617 491 0053)

SERVATOR RANGE of composting toilets, The Old Rectory, Easkey, Co. Sligo. (tel. 00 353 9649181)
Email:postmaster@laci.demon.co.uk
WWW: http://www.greenbiz.co.uk

Contact: Robers Forrester

UK/Ireland agent for Servator Swedish manufacturers of composting toilets including 12 volt models.

ERNST H2NO! Waterless Urinal GLOBEMALL LTD Jonathan Marland 1 Woodbridge Road, Ipswich, IP4 2EA
(tel: 01473 259232　fax: 01473 286285)

Whilst the concept of ''waterless urinals'' seems new in this country, the Ernst company in Switzerland launched its waterless urinals just over 100 years ago in 1884, and there are literally hundreds of thousands of their products installed in various countries. Here in the UK we have installations with Borough & County Councils, Offices, Supermarkets, Universities and Pubs.

Listing supplied by the Green Building Press, extracted from 'GreenPro' the building products and services for greener specification database. At present GreenPro lists over 600 environmental choice building products and services available throughout the UK and is growing in size daily. The database is produced in collaboration with the Association for Environment-Conscious Building (AECB).
For more information on access to this database contact Keith Hall on 01559 370908 or email buildgreen@aol.com

1. Fertile Waste - Managing your Domestic Sewage (P Harper & D Thorpe, Centre for Alternative Technology, Machynlleth 1994)

2. Dictionary of Environmental Science & Technology (Andrew Porteus, John Wiley & Sons, Chichester 1992)

3. Environmental Chemistry, 2nd Edition (Peter O'Neill, Chapman & Hall, London 1993)

4. The World Environment 1972-1992 - Two Decades of Challenge (M. K. Tolba & O. A. El-Kholy (eds), Chapman & Hall, London, for The United Nations Environment Programme 1992)

5. Hazardous Waste & Human Health (The British Medical Association, Oxford University Press, Oxford 1991)

6. Design with energy - The conservation and use of energy in buildings (J Littler & R Thomas, Cambridge University Press, Cambridge 1984)

7. Ecological Building Factpack (R Pocock & B Gaylard, Tangent Design Books, Leicester 1992)

8. Places of the Soul - Architecture and Design as a Healing Art (Christopher Day, The Aquarian Press/Thorsons, Wellingborough 1990)

9. Eco-Renovation: The ecological home improvement guide (Edward Harland, Green Books, Dartington 1993)

10. The Building Regulations - Explained & Illustrated, 9th Edition (V Powell-Smith & M J Billington, Blackwell Scientific, Oxford 1992)

11. The Building Design EASIBRIEF (Henry Haverstock, Morgan-Grampian (Construction Press) Ltd 1993)

12. The Green Construction Handbook - Going Green - A Manual for Clients and Construction Professionals (Ove Arup & Partners, J T Design Build Ltd, Bristol 1993)

13. Environmental Standard - Homes for a Greener World, BR 278 (Josephine J Prior & Paul B Bartlett, Building Research Establishment, Garston 1995)

14. The UK Environment (Department of the Environment, HMSO, London 1992)

15. Sewage sludge incineration processes under 1 tonne an hour - Secretary of State's Guidance, PG 5/5(91) (Department of the Environment, HMSO, London 1991)

16. Waste Disposal & Recycling - Sewage Sludge Incineration Process Guidance Note, IPR 5/11 (Department of the Environment, HMSO, London 1992)

17. Code of Practice for Agricultural Use of Sewage Sludge (Department of the Environment, HMSO, London 1989)

18. Gardening Which? July 1995 (Consumers' Association, Peterborough 1995)

19. The Toilet Papers - Recycling Waste and Conserving Water (Sim Van der Ryn, Ecological Design Press, California 1995)

20. From (19) above, a quote from Dr. Vivian Poore, originally quoted in Conservancy or Dry Sanitation versus Water Carriage, J Donkin (1906)

21. Sanitation without Water, revised & enlarged edition (Uno Winblad & Wen Kilama, Macmillan, London 1985)

22. Small Scale Sanitation, Bulletin No. 8 (Sandy Cairncross, The Ross Institute, London 1988)

23. Sewage Treatment using Constructed Reed Beds - a brief introduction, Tipsheet 3 (Centre for Alternative Technology, Machynlleth)

24. The Design of Septic Tanks and Aqua-Privies, OBN 187 (John Pickford, Building Research Establishment, Garston 1980)

25. Disposal of Domestic Effluents to the Ground, OBN 195 (R F Carroll, Building Research Establishment, Garston 1991)

26. Health Aspects of Latrine Construction, OBN 196 (R F Carroll, Building Research Establishment, Garston 1991)

27. Building For A Future, Spring/Summer 1995 vol.5 no.1 (Association for Environment Conscious Building, Coaley 1995)

28. Acheiving Zero Dioxin - An emergency Strategy for Dioxin Elimination (Greenpeace International, London) 1994

29. Chlorine-Free Vol.3 No.1 (Greenpeace International, London) 1994

30. Social Trends, HMSO, 1995

31. Jan McHarry; 'Reuse, Repair, Recycle' (Gaia Books, 1993)

32. Pearce, D.W., Markandya, A. and Barbier, E.B. (1989) Blueprint for a Green Economy (Blueprint 1), Earthscan, London.

33. Mann, H.T., Williamson, D., 'Water Treatment and Sanitation' (Intermediate Technology Publications, 1993)

Carpets and Floorcoverings 15

15.1 Scope of this Chapter

This chapter looks at the environmental impacts of the main floorcovering materials available on the market, covering carpets and underlays, vinyl, lino, wood, cork, rubber and stone. Ceramic tiles and in-situ resin, latex and epoxy floors are not covered.

Varnishes and glues, used in association with floorcoverings, have not been covered in any detail.

15.2 Introduction

The main issues relating to floorcoverings are the environmental impacts of the manufacture of synthetic sheets, fibres, rubbers and foams, and the health issues linked with the emission of Volatile Organic Compounds (VOCs) from synthetic floorcoverings. The issue of VOC emissions from carpets and its links with Sick Building Syndrome has recieved a lot of attention in the USA,[52] although it has so far avoided the media spotlight in the UK.

15.2.1 Synthetic vs. "Natural" Floorcoverings

"Natural" floorcoverings may have a smaller environmental impact than their synthetic counterparts, but tend to be more expensive due to long distance transport (eg. plant fibres from the sub-tropics, or wool sourced from New Zealand), high processing costs[24] and their use almost exclusively as 'high quality' floorings, with few low cost alternatives.

The notable exception is Linoleum which, according to our research, is comparable in price to its synthetic counterpart, Vinyl (PVC). Vinyl is currently the market leader for 'smooth' floor coverings, taking over 80% of vinyl and lino sales[29] although Linoleum (made from linseed oil, cork, wood flour and powdered limestone, pressed onto a hessian backing[35]) is starting to make a comeback due to the current popularity of 'natural' products.[28] Besides the environmental benefits, lino is also softer in feel and warmer on the feet than PVC.[27] Linoleum also gives off linseed oil vapour, which kills bacteria without being toxic to humans.[26]

Both Lino and vinyl (PVC) floorings are popular, taking 70% of smooth floorcoverings sales, mainly because they are both cheap, easy to clean and available in a wide variety of colours and designs.

During our research, we found that many retailers and wholesalers tend to confuse linoleum with vinyl, treating them as one and the same. It is therefore important to specify linoleum as opposed to vinyl/PVC.

The lower price end of the market is dominated by synthetics[28] which are mass produced cheaply, and used in low cost floorings such as needlepunch carpet tiles and vinyl. The low price of synthetic products masks the hidden costs of environmental damage during manufacture, health hazards and the use of non-renewable resources.

15.2.2 Durability and Recycling

The durability of a floorcovering is important in determining its environmental impact. For frequently refurbished offices this may be considered less important, as the floorcovering may be removed before it is worn out.

Stone, ceramic tile and hardwood floors are the most durable, due to the nature of the material, lasting for tens to hundreds of years. Wool and acrylic tend to be used in 'high quality' carpets, which are usually the most durable. Synthetic carpet tends to be less durable, not due to the materials themselves, but because they are used mainly in the manufacture of low cost carpets which are often not built to last, thus having a greater environmental impact in the long run.[24] Similarly, latex (natural rubber) backing and underlay is reported to be more durable than many synthetic rubbers. For example, polyurethane foam tends to disintegrate to powder after a few years, particularly if any liquid is spilled onto the carpet.[9]

Felt, the traditional carpet underlay, can be manufactured from recycled fabric.[9] Felt underlay is the third largest UK market for reclaimed fibres, accounting for 5,000 tonnes per year. The main advantage of this is the energy savings; 100% recycled wool products saves 50% energy compared with new wool, because it does not need transporting (often from Australasia), scouring, carbonising or dyeing. Recycling also reduces freshwater consumption and effluent production.[37]

Wool, wood, cork, stone and possibly linoleum floors are all potentially recyclable,[24] and reclaimed stone and ceramic floorings are available on the market.

The recycling of synthetic floorcoverings is technically more compex and rarely carried out.

15.3 Best Buys

15.3.2 Carpet

The best 'green' buy is wool carpet with hessian backing, using a recycled felt underlay, or other "natural" products listed in the Alternatives section. These tend to be in the mid- to high-cost end of the market.

The low cost end of the market is dominated by synthetic carpets with synthetic foam or PVC backing and usually fixed using solvent based glues. All of these have impacts on both the environment and health. The difference between most of the fibre materials is marginal, except for Nylon which appears to have the highest environmental impact. Latex backings are generally more benign than synthetics, and have higher durability, as described above. The additional impact of using a foam depends on the blowing agent used. These could be HCFC/HFCs or other ozone depleting / greenhouse gases, although more environmentally benign blowing agents are now coming into widespread use.

The 'green' option for fitting carpet with regard to occupant health is to use grippers or tacks rather than solvent based adhesives. For carpet tiles and lightweight carpeting for which these are not appropriate, non-solvent based adhesives are available. This subject is planned for a future issue of the Green Building Digest magazine.

15.3.1 Smooth coverings

At the low to medium cost end of the market, this is basically a choice between vinyl, linoleum, rubber or the cheaper cork and parquet flooring.

Linoleum is by far the 'greenest' of the lino/vinyl type floors, performing as well as PVC in terms of durability and appearance and only marginally more expensive, while avoiding the environmental costs of vinyl (PVC). Cork is also a 'green' option, a renewable resource produced in the Iberian peninsula by small firms.[24] Cork is susceptible to abrasion damage, but is otherwise highly durable. It has been in use as a flooring material for more than 100 years.[34] and it's use "is to be encouraged".[24]

Where rubber floorings are to be used, latex (natural) rather than synthetic rubber, has a higher durability (see previous page) and lower environmental impact. Latex however, has poor resistance to spills of oils, solvents and oxidants.

The mid- to high-cost end of the market is dominated by 'green' products - stone, ceramic, parquet and cork. Stone and ceramic are non-renewable resources but have extremely high durability and can be reused almost indefinately. Reclaimed ceramics and stone are available on the market, and their use avoids the environmental impacts of extraction incurred by using a virgin resource. Parquet floors are manufactured from a potentially renewable resource, but many sources, particularly tropical hardwoods, are not sustainably managed. Specifiers may wish to check timber certification to ensure that the timber is from a sustainably managed source. Timber certification is given extensive coverage in chapter 7, and a brief summary is given in this chapter.

The Unit Price Multiplier

The unit price multiplier is derived from the average of the price range for a particular product. Extremely high cost 'luxury' items have been omitted from the calculation so as not to distort the result. However, some still represent an enormous range - eg; parquet flooring can cost anything from £10 to over £50 and stone can cost from £50 to over £200 per square metre.[51]

Carpet pricing methods are not easily applicable to this report, as they depend on the style, colour, weight and brand of carpet as well as the material from which they are manufactured. Many popular carpet brands are an 80:20 mix of wool and synthetic; it is therefore almost impossible to give a unit price for each individual carpet material. The unit price multipliers for carpet fibres and backings are therefore only a very rough guide.

Best Buys

Carpet

Best Buy:	Wool
Avoid:	Nylon

Underlay

Best Buy:	Hessian/Felt
Avoid:	Synthetic Foams

Smooth Coverings

Best Buy:	Linoleum, Cork, Timber, Stone.
Second Choice:	Ceramic Tile, Latex Tile
Avoid:	Vinyl/PVC

Fixings

Best Buy	Grippers/Tacks
Avoid	Solvent-Based Glues

| | £ | Production | | | | | | | | Use | | | | ALERT! |
	Unit Price Multiplier	Energy Use	Resource Depletion (bio)	Resource Depletion (non-bio)	Global Warming	Ozone Depletion	Toxics	Acid Rain	Other	Durability/Maintenance	Recycling/Reuse/Disposal	Health	Other	ALERT!
Carpet Fibre/Pile														
Wool	1.6	·					●			·				
Nylon	0.6	●		●	●	●	●	●		·	●	·	·	
Polyester	-	●		●	●		●	·		·	●		·	
Polypropylene	-	●		●	·		●	·		·	●		·	
Acrylic	0.8	●		●	·		●	·		·	●	·	·	
Carpet Backing/Underlay														
Hessian	-	·												
Felt (Recycled)	-	·												
Polyethylene	-	●		●	·		●	·			●		·	
Polyurethane	-	●		●	·		●			●	·	●	·	
Polyurethane Foam	-	●		·	●	●	●	●		●	●	●	·	
Latex	-	·		·			·			·	·	·	·	
Latex Foam	-	·		·	●	●	·			·	·	·	·	
Smooth Floorcoverings														
PVC/Vinyl	0.6	●		●	·		●	·			●	●	·	Greenpeace Campaign
Linoleum	0.6	·		·										
Cork	1	·	·				·			·		●		
Wood/Parquet	1-2.4	·	●									·		
Stone	8.1	·		●								·		
Latex Tile & Sheet	-	·		·			·			·	·	·	·	
Butadine/Styrene Tile	1	●		●	·		●	·		·	●	·	·	

A Note on Methodology

Carpets are constructed from a number of different materials - the pile, which may be a composite of 2 or more types of fibre, the backing and the underlay. This digest has looked separately at each of the common componants of carpets, grouped as pile/fibres, backings and underlay, rather then looking at specific brands.

15.4 Product Analysis

15.4.1 Synthetic Fibres, Foams & Sheeting

All of the synthetic fibres, sheets, tiles and foams listed in this report are products of the petro-chemical industry, [11] and therefore have similar impacts. This introductory section outlines the general impacts of synthetic products to save repetition of information in each material section. The specific impacts of each material are dealt with individually.

Production

Energy Use

Plastic polymers are produced using high energy processes, using oil or gas as raw materials, which themselves have a high embodied energy.[15]

Resource Depletion

Unless otherwise stated, Oil is the raw material for all the synthetic materials listed in this report. This is a non-renewable resource. Some plastics are manufactured from vegetable oils, but we found no evidence of these being used in the production of the fibres, sheets or foams used in carpet manufacture.

Global Warming

Petrochemicals manufacture is a major source of NOx, CO_2, Methane and other 'greenhouse' gases.[19]

Acid Rain

Petrochemicals refining is a major source of SO2 and NOx, the gases responsible for acid deposition.[19,15]

Toxics

The petrochemicals industry is responsible for over half of all emissions of toxics to the environment, releasing a cocktail of organic and inorganic chemicals to air, land and water. The most important of these are particulates, organic chemicals, heavy metals and scrubber effluents.[19]

Other

The extraction, transport and refining of oil can have enormous localised environmental impacts,[15] as illustrated by tanker accidents such as the Exxon Valdez and Braer spills, and the environmental degradation of Ogoniland, Nigeria.

Use

Recycling/Reuse/Disposal

Thermoplastics, which can be melted and reformed, are potentially recyclable, but the wide variety of plastics present in waste make separation and recycling an expensive and complex process. Currently, post-consumer recycling of plastics is negligable.[17] This is a particular problem with plastics used in carpeting, for which a variety of synthetic and natural materials are combined in the same product. Thermoset plastics cannot be re-moulded by heating and are therefore not recyclable.

Health Hazards

Synthetic carpets are a recognised source of Volatile Organic Compounds (VOCs), some from the pile, but mainly from backing and adhesives. The large surface area presented by carpet gives the potential to seriously affect indoor air quality,[6] and 'sick building syndrome' is often attributed to carpeting and other flooring materials. Carpets release a number of VOCs, the ones of most concern being Styrene, 4-Phenylcyclohexane (4-PCH - the source of 'new carpet' odour), 4 Ethynylcyclohexane (4-ECH) and formaldehyde.[6] The main source of these are the styrene-butadiene latex adhesives commonly used to bind the secondary backing of carpets.[6] Formaldehyde is also released by the carpet backings themselves although its use is being phased out by many manufacturers. The US Carpet and Rug Institute (CRI) claim that formaldehyde has not been used in the manufacture of

---

Formaldehyde

Formaldehyde is the simplest aldehyde; a colourless, pungent and very reactive reagent.
Although highly toxic, it thought only to be a problem if very large areas of formaldehyde releasing material, such as flooring, solvents or chipboard is present, with no ventilation.[14] Formaldehyde is highly toxic if inhaled, a powerful skin irritant, and a suspected human carcinogen.[12, 16] At low concentrations, it causes headaches, and irritation to the eyes and throat, and may cause an allergic reaction in susceptible individuals.[12, 16]
The concentrations of formaldehyde are very low, even in new carpet[7, 16] and problems can be avoided with adequate ventilation.[16]

Volatile Organic Chemicals and Ozone

Emissions of VOCs from synthetic carpets and backings are enhanced by the presence of ozone. Ozone levels comparable to city pollution levels were found to increase carpet emissions of formaldehyde by a factor of 3, and acetaldehydes by a factor of 20. This is due to reaction of ozone, a powerful oxidant, with compounds in the carpet.[6]

carpets in the US (where 50% of carpets are manufactured[28]) for over 12 years,[7] although this claim is contradicted by the findings of research carried out for the US Consumer Safety Commission, which detected formaldehyde in a number of the new carpets analysed.[52] With the exception of formaldehyde, a known carcinogen, only limited data is available regarding the irritancy and toxicity of VOCs at low concentrations.

The most extreme symptoms of carpet-related sick building syndrome are likely to be felt on visits to carpet departments and warehouses, where people with chemical sensitivity can experience depression, burning eyes, headaches, sore throat, irritability and/or palpitations. This is mainly due to formaldhyde in fabric treatments and underlay.[8] The problem is generally far less noticable in normal office or home environments where levels of VOCs from carpet are low, and the concentrations decrease rapidly after installation,[7] to less than 1ppbv or less after one week. An extensive study of carpet emissions carried out for the US Consumer Safety Commision, found that under normal conditions, carpets emit formaldehyde and other volatiles at a rate resulting in concentrations below the irritance threshold of 70ppbv, but combined with other sources of VOCs around the home, new carpet emissions could result in concentrations above threshold limit values, particularly in poorly ventilated buildings. The report concludes that too little information is available regarding the health effects of many of the VOCs emitted by carpets, but it is unlikely that they will have any significant effects.[52] In most cases, sick building syndrome is thought to relate more to air *quality* and *comfort*, rather than a specific health hazard.[32]

15.4.2 Carpet Fibres

(a) Wool

Wool is the traditional material for carpet fibres, and is still very popular due to its attributes - softness, durability, and ability to provide a pile with 'springiness'. Wool also dyes well and gives good thermal and sound insulation. The main drawback is its high cost[28] - a cheap wool carpet is comparable in cost to an expensive synthetic carpet.

Production

Resource Use

Wool is a natural, renewable resource. Most of the wool for the UK carpet industry is sourced from Scotland and West Yorkshire where the highest density of sheep farming is found. The carpet industry is traditionally concentrated in these areas.[28]

Toxics

Organophosphates, used in sheep dips, have been linked to a range of physical illness, depression and mood swings. The suicide rate amongst sheep farmers is double the national average,[41, 43] which has lead to tight controls on their use since April 1995.[42] Organophosphate released into rivers by careless disposal of sheep dip effluent and fleece scouring plants is suspected to be partially responsible for the reduction of fish stocks in UK rivers.[41]

Wool production in New Zealand, Australia, South Africa and Uraguay (the worlds major producers) has a relatively trouble free environmental record - indeed, southern hemisphere producers are actively seeking ways to cut the chemical input in their sheep dips.[41] However, carpet producers will tend to source their wool from the cheapest supply at any one time. As a result, it is almost impossible to determine the origin of the wool in a particular carpet and therefore impossible to make an informed decision

Dyes

Most fabric dyes are synthetic, being derived from petrochemicals.[46]

The mordants used to fix many synthetic dyes contain the heavy metals lead, tin, chrome and titanium, and are often discharged into rivers where they are highly toxic to aquatic organisms.[47]

Synthetic 'reactive' dyes, which produce the bright and fast colours on natural fibres, are not totally fixed to the fibres, leaving 10-50% of the dye in the effluent. Spent reactive dyes are not removed by conventional sewage treatment processes, resulting in colour pollution of rivers.[48]

"Natural" fibres can be dyed using natural plant dyes, which are fixed using salt and lemon rather than mordants.[49]

Specifying undyed, unbleached wool or cotton carpet is another way of avoiding the environmental impact of synthetic dyes. Avoiding synthetic dyes is not possible with synthetic fibres, but low impact synthetic dyes are available which do not contain heavy metals, use less water and energy than normal synthetic dyes, and can result in up to 75% less unused dye in the effluent.[49]

regarding organophosphates when specifying wool carpets.

Many of the dyes, bleaches, moth proofers and fire retardants used in wool treatment produce toxic wastes or by-products (See Dyes section below). Wool, however, requires far fewer chemicals to treat it compared to other fibres.[41]

Use

Health

Wool is reported to be healthier in the home than its synthetic counterparts,[23] although ''natural'' fibre carpets tend to harbour allergens, dust mites and fleas to a greater extent than synthetic fibres.

Recycling/Reuse/Disposal

Wool is a natural, biodegradable material.

Fire

Polyurethane and polyester release toluene diisocyanates when burned. Polyurethane catches fire easily and burns rapidly giving off a dense black smoke containing cyanides. Since Mach 1990, polyurethane foams used in furnishings must be combustion modified to ignite and burn less readily.[10] PVC burns to release toxins such as hydrogen chloride,[8] dioxins and phosgene gas.[9] Both Acrylic and Polyurethane release hydrogen cyanide gas when burned.[8]

Natural fibres and latex also release combustion products with similar toxicity to burning wood,[12] but at a much slower rate than their synthetic alternatives, allowing a longer time for escape before being overcome by toxic smoke.

Durability

Wool wears well, and wool carpets are generally manufactured to high quality specifications, as they are aimed at the higher cost, higher quality end of the market.

(b) Nylon

Nylon is the long-standing alternative to wool, often used in blends of 20% nylon, 80% wool, and adds resilience [28]

Production

Energy Use & Resource Depletion;

(See 'Synthetic Fibres, Foams and Sheets' section p.198)

Global Warming

10% of the annual increase in atmospheric NOx and more than half the UK production of NOx originates from nylon manufacture.[13, 41] Nitrous oxide is the third most important greenhouse gas, after CO_2 and Methane.

Ozone Depletion

NOx also contributes to ozone depletion.[13]

Acid Rain

NOx in the atmosphere forms acid rain.

Other

(See 'Synthetic Fibres, Foams and Sheets' section p.196)

The technology is available to reduce emissions, and some manufacturers burn off nitrous oxides and other gases before emitting exhaust gases from nylon manufacturing plants.[13]

Use

Health Hazards

Nylon is generally considered safe during use.[10]

Recyling/Biodegradability

Nylon is persistent in the environment,[24] and releases toxic fumes when incinerated.

(c) Polyester

Polyester is used only minimally in carpet manufacture, with carpets containing over 50% polyester making up no more than 1% of the sector.[28]

Production

Energy Use, Resource Depletion, Global Warming & Acid Rain;

(See 'Synthetic Fibres, Foams and Sheets' section p.198)

Toxics

Although associated with toxic emissions, polyester manufacture has a relatively small impact when compared with PVC.[20] Polyester dyeing requires the use of dye carriers, which allow the dye to penetrate the fibre. Many of these are known carcinogens[47] (See dyes section on facing page).

Other

(See 'Synthetic Fibres, Foams and Sheets' section p.198)

Use

Recycling/Reuse/Disposal

(See 'Synthetic Fibres, Foams and Sheets' section p.198)

Health Hazards

Polyester fibres are reported to cause eye and respiratory tract irritation, and acute skin rashes.[10]

(d) Polypropylene

The use of polypropylene is growing, and currently accounts for almost 20% of the UK tufted carpet output.[28]

Production

Energy Use

As with most plastics, Polypropylene has a high embodied energy (100Mj/kg).[11]

Resource Depletion, Global Warming, Acid Rain, Toxics, Other;

(See 'Synthetic Fibres, Foams and Sheets' section p.198)

Use

<u>Health Hazards</u>

Polypropylene is a low toxicity polymer.[16]

<u>Recycling/Reuse/Disposal</u>

Polypropylene is highly persistent in the environment,[16] making disposal a problem. 7% of UK production is currently recycled.[9]

(e) Polyurethane
(See Foam and Underlay)

(f) Acrylic

Used mainly for synthetic Axminsters and Wiltons, acrylic is not favoured by the tufted sector. Imported acrylic Wiltons account for 75% of the UK Wilton market in 1994.[28]

<u>Energy Use, Resource Depletion, Global Warming, Acid Rain, Other, Biodegradability, Health</u>

(See 'Synthetic Fibres, Foams and Sheets' section p.198)

<u>Toxics</u>

Acrylonitrile, the colourless liquid used to make acrylic resins, is a suspected carcinogen

and has been known to cause breathing difficulties, headaches and nausea.[9,10]

15.4.3 Backings and Underlay

(a) Hessian

The traditional carpet backing, made from Jute plant fibres. Hessian is environmentally the best option for carpet backing, as it is produced from a renewable resource, requires minimal processing, is highly durable, non-toxic and biodegradable.

(b) Felt

Felt, the traditional carpet underlay, can be manufactured from recycled fabric.[9] Felt underlay is the third largest UK market for reclaimed fibres, accounting for 5,000 tonnes per year. The main advantage of this is the energy savings; 100% recycled wool products saves 50% energy compared with new wool, because it does not need transporting (often from Australasia), scouring, carbonising or dyeing. Recycling also reduces freshwater consumption and effluent production.[37]

(c) Polyethylene

Used as an underlay for cork and parquet flooring as protection against sub-floor moisture.

Production

<u>Energy Use, Resource Depletion, Global Warming, Ozone Depletion, Toxics, Acid Rain, Other;</u>

(See 'Synthetic Fibres, Foams and Sheets' section p.198)

Use

<u>Health Hazards</u>

Polyetheylene is a low toxicity plastic.[16]

<u>Recycling/Reuse/Disposal</u>

Polyethylene is a thermoplastic, and potentially recyclable. Many polyethylene products now have chalk mixed in to promote their breakdown into small fragments when buried.[16]

(d) Acrylic (Iuctite)

Luctite, non-woven acrylic fabric, is used as carpet backing.[10] - (see acrylic fibre for impacts)

15.4.5 Synthetic Foams and Rubbers - General

<u>Health Hazard</u>

Cadmium impurities in rubber carpet backings are thought to be the main source of cadmium dust contamination in the indoor environment. This has not been quantified as a significant health risk, and the possibility of sub-clinical effects on childrens intellectual development remains controversial.[12]

Blowing Agents

Materials such as Latex and Polyurethane require the use of blowing agents to form foams.

Ozone destroying CFCs, used as blowing agents in many older processes, are gradually being replaced by HFCs and HCFCs. These in turn are being phased out due to their greenhouse gas potential, which is 3200 times that of carbon dioxide.

Alternative processes now use carbon dioxide, ammonia and other organic solvents which are considered less environmentally damaging, although by no means benign.[4]

The most common blowing agent for polyurethane foam is dichloromethane, a chlorinated hydrocarbon.[38] Others include solvents such as Methylene Chloride, Acetone, CFC's and Isopentane. To permit easier bubble formation, blowing also uses surfactants[1] which can have a detrimental effect on aquatic fauna when the effluent is released into watercourses.

(a) Polyurethane Foam

Production

Toxics

Polyurethane is manufactured from Polyol and Toluene Diisocyanate (TDI), using an amine catalyst. TDI has been linked to Reactive Airways Dysfunction Syndrome (RADS) in workers exposed to high doses.[2] The symptoms, similar to asthma, can be brought on by a single exposure and have been recorded as lasting up to 5 years.[2,3] Once sensitised to TDI, exposure to even extremely low concentrations can lead to a severely disabling reaction.[12]

A by-product of polyurethane production is the highly toxic phosgene gas.[9]

Energy Use, Resource depletion, Acid rain;

See 'Synthetic Fibres, Foams and Sheets' section p.198

Ozone Depletion & Global Warming

See 'Blowing Agents' Section below, and 'Synthetic Fibres, Foams and Sheets' section p.198

Use

Health Hazard

Polyurethane foams are significant sources of formaldehyde.[8] Polyurethane causes cancer in laboratory animals, but it is unlikely that people would absorb harmful amounts of this substance from flooring materials.[9] Hydrogen cyanide gas is released when polyurethane is burned.[8] Also see 'Synthetic Rubbers and Foams' & 'Synthetic Sheets (etc)' sections p.198.

Recyling/Reuse/Disposal

Polyurethane is a thermoset plastic, which means that it cannot be remelted or reformed, and is therefore not recyclable.[17]

Durability

Polyurethane foam is not durable, and tends to disintegrate to powder after a few years, particularly if any liquid is spilled onto the carpet.[9]

(b) Butadine-styrene Co-polymers

A synthetic material widely used in floor tiles.[25]

Production

Energy Use, Resource Depletion, Acid Rain, Other;

(See 'Synthetic Fibres, Foams and Sheets' section p.198)

Toxics

Synthetic rubbers manufacture often involves the use of hazardous chemicals - both Butadine and Styrene are possible carcinogens.[12] Once combined, it is thought that there is no significant risk to construction operatives or to building users from these synthetic rubbers.[25]

Ozone Depletion & Global Warming

See Blowing Agents section below & 'Synthetic Fibres, Foams and Sheets' section,p.198.

Use

Recycling/Reuse/Disposal

Synthetic rubbers can be broken down and reclaimed for low-specification products.[25]

Health Hazards

See 'Synthetic Foams and Rubbers' section, p.198

(c) Latex (Natural) Rubber & Foam

Production

Latex is a natural rubber product derived from trees, historically used extensively for flooring material and underlay. Latex requires coagulation, and the addition of vulcanising agents, accelerators and fillers to convert it to a usable material, but the use of natural rubber is still to be encouraged over its synthetic counterparts as it is a renewable resource.[25]

Ozone Depletion & Global Warming

Latex foams should be treated with more caution than rubber sheet or tile, as they require the use of blowing agents such as HFCs and ammonia, most of which are environmentally damaging (See Blowing Agents Section below).

Use

Health Hazard

Latex is not anticipated to pose any significant toxic hazard to residents, and stripping during renovation or demolition poses no threat to workers, unless mineral fibres are incorporated in the matrix.[12]

Durability

Durability and flexibility properties are superior in many ways to those of its synthetic substitutes,[23] although natural rubber is limited in its applications as it lacks resistance to oxidation, oils and solvents.[25]

Recycling/Reuse/Disposal

The environmental impacts of dumping and incineration of latex foam have not been evaluated.[12]

15.4.6 Smooth Floor Coverings

Vinyl sheets and tile, linoleum sheet and tile, cork, wood (parquet, mozaic, woodstrip) & stone.

(a) PVC (Vinyl)

Production

Energy Use

The production of ethylene and chlorine, the raw materials of PVC, is extremely energy intensive. However, PVC has one of the lowest embodied energies of plastic materials (68Mj kg⁻¹).[11]

Toxics

PVC is manufactured from the vinyl chloride monomer

and ethylene dichloride, both of which are known carcinogens and powerful irritants.[9,15] PVC also contains a wide range of additives such as fungicides, pigments, plasticisers and heavy metals, which adds to the toxic waste production.[12,17]

High levels of dioxins have been found in the environment around PVC production plants,[17] and the waste sludge from PVC production going to landfill has been found to contain significant levels of dioxins and other highly toxic organic compounds.[17,18] PVC production is top of Her Majesty's Inspectorate of Pollution list of toxic emissions to air, water and land.[17]

Resource Depletion

Alongside the use of oil as the primary raw material, rock salt used to provide chlorine for PVC production, is mined in significant quantities.

Global Warming, Ozone Depletion, Acid Rain, Other;
(See 'Synthetic Fibres, Foams and Sheets' section p.198)

Use

Health Hazards

PVC flooring releases high concentrations of plasticiser, an additive used to improve flexibility, which contributes to 'sick building syndrome'.[17] The main plasticiser used in vinyl flooring is butyl benzyl phthalate, which releases benzyl- and benzal chloride.[31] PVC also releases unreacted vinyl chloride monomer left over from the manufacturing process, but in very small amounts.[14] Release problems are increased during heating - for instance, when PVC tiles or sheets are soldered together by heating with a wire.[12]

Recycling/Reuse/Disposal

As with most plastics, PVC is persistent in the environment. It is potentially recyclable, for use in low-grade thick walled items such as park benches and fence posts. Post consumer recycling of PVC is negligable.[17] Incineration is not recommended as it produces toxins such as dioxins and leaves 90% of the original mass in the form of waste salts with the release of only 10% of PVCs embodied energy.[50] The possibility of leaching plasticisers and heavy metal stabilisers means that landfilling is also a less than safe disposal option.[18]

Alert

The environmental group Greenpeace is campaigning world-wide for an end to all major industrial chlorine chemistry, including the manufacture of PVC, due to its toxic effects.

(b) Cork

Production

Energy Use

Cork requires very little energy in production (mainly in the processes of boiling the virgin material, and baking agglomerated cork). Cork is grown in the Meditteranean,[34] and so the energy costs of transport must be considered, although these are very low in relation to the more weighty stone or wood.

Resource Use

Cork is a renewable material and environmentally benign. It is the bark of the evergreen oak *Quercus suber,* which can be stripped every 10-12 years without harming the tree[22] although they are more susceptible to injury until the outer bark has regenerated.[34] According to recent report by Environmental Building News, the cork forests are treasured, and sustainably managed. Most producing countries regulate the frequency of bark harvesting in

Cleaning and Maintenance

Smooth floor coverings are easy to clean by manual brushing and mopping. The environmental impacts are low, although depend on the detergent used. Additional effort and impact is incurred with wooden floors, which may require varnishing and polishing. Nevertheless, smooth, easily cleanable surfaces pose far fewer maintenance and air quality problems than carpet.[12]

Levels of airborne dust and bacteria in room air measure highest in carpeted areas; old carpets are associated with the highest readings.[39] Carpet cleaning requires a vacuum cleaner, with associated energy and resource use in manufacture and use. (reviewed in Ethical Consumer Magazine, no.26[40]). Alternatively, manual devices such as carpet sweepers or dust pan and brush can be used.

Carpet shampoos often contain Perchloroethylene, which causes long-term damage to vital organs and is a suspected carcinogen, Napthalene, a suspected carcinogen[9] or Ammonia, which can cause irritation to the respiratory system. It is also a threat to aquatic life if effluents enter waterways.[8] Alternatives to conventional carpet cleaners include:

-1 part washing up liquid, 2 parts boiling water whipped with a beater and applied to the carpet[9]

-Sprinkle bicarbonate of soda, leave overnight, and hoover off. This also acts as a deodoriser[9]

A major drawback of natural, coarse fibre products such as sisal and grass matting is that they are regarded as being difficult to clean. Synthetic carpets are also regarded as easier to maintain than wool - as one carpet retailer put it ''which is easier to wash - a pure wool sweater, or one made from man made fibres?''.

order to protect the trees, and in Portugal it is illegal to fell trees other than for essential thinning and removal of old, non-productive trees.[34]

Durability

Cork can withstand extremely high pressure without permenant deformation, is light weight, durable, non-flammable[23] resistant to moisture damage and decay [(34)] and warm on the feet,[23] although it has poor abrasion resistance and is vulnerable to high heels.[22, 25]

Timber Sources and Certification

(See Timber chapter for further detail.)

Using locally grown timbers, (usually indicated by the FICGB Woodmark), saves on transport energy, and British forestry is generally well managed.[44] Scandinavian wood (eg: Kahrs parquet) has the least far to come of the imported timbers, whereas tropical timber has the furthest to travel, as well as being associated with the worst forestry practices.

Wood Certification is carried out by 4 main organisations; Scientific Certification Systems Ltd, Smart Wood (Rainforest Alliance), SGS Forestry and the Soil Association which are in turn monitered and accredited by the Forest Stewardship Council (FSC). A comprehensive list of certification organisations can be found in the Timber chapter.

Westco's Rhodesian Teak parquet flooring is independently certified by SGS Forestry as coming from a sustainably managed source - perhaps the first certified flooring in the world. Other manufacturers are sure to follow their lead, as the major DIY stores are committed to achieving total independent certification of timber products by the year 2000.[29]

Use

Health Hazards

Cork flooring is formed from ground cork granules (leftover from bottle stopper manufacture) which are mixed with a binder, moulded into blocks and baked - a process which produces almost no waste.[34] Until the 1980s, urea-formaldehyde glues were used. These are reported to have been replaced with "all natural protein binders", although there is evidence that less benign binders such as urea melamine (a mixture of urea formaldehyde and melamine formaldehyde) and polyurethane are still used for cork agglomoration by some major manufacturers.[34]

In summary, cork products are relatively benign with regard to resource use, but there may be cause for concern regarding occupant health. Also, beware of vinyl-cork composite floor tiles, which have a vinyl (PVC) backing or surface coating. This increases abrasion resistance, but also significantly increases the environmental impact of manufacture. If safe binders are used, natural cork products "may become the ideal flooring for many applications".[34]

(c) Linoleum (Lino)

Production

Resource Use

Linoleum is made from a linseed oil and natural resin 'cement' mixed with cork, wood flour and powdered limestone, which is pressed onto a woven hessian backing.[35]

Most constituents are natural and renewable, although non-renewables such as bitumen felt backing may sometimes be used.[25, 26]

Health Hazards

To avoid the health hazards posed by solvent based adhesives, Lino should be installed with a lignin paste or other 'safe' adhesive.[27]

(d) Wood Floors

Although essentially a 'green' option, being a natural, potentially renewable, durable and eventually biodegradable material, there are concerns regarding wood floorings. These are namely the use of non-sustainable tropical hardwoods and solvent based adhesives and varnishes.[25]

Production

Resource Depletion

Hardwoods are the most resistant to wear and have found wide use, but many of these are sourced from non renewable rainforest sources. The more decorative types such as parquet should therefore be avoided unless the source can be guaranteed.[25] Wood-block, strip and mozaic floors are to be preferred as they tend to use softwoods or temperate hardwood.[25] Nevertheless, purchasers may wish to check that the source of wood is certified renewable (See Chapter 7).

Wood floor finishings from acceptable sources are to be encouraged as a renewable resource.[25]

Use

Toxics

Fine wood dust, released during installation/maintenance is a suspected carcinogen, and tropical wood dusts may have respiratory effects.[14] It should also be noted that the fumes given off by burning wood are similar in toxicity to some synthetic materials.

Glues and resins used to make pre-formed parquet/mozaic floors, may break down to produce irritant volatiles.

Durability

Timber floors are extremely durable.

Recycling/Reuse/Disposal
Wood floors are reusable and biodegradable.

VARNISHES

Wood and cork flooring is often finished with a varnish, which can diminish the otherwise good environmental credentials of the product.

Polyurethane

Consists partially reacted polyisocyanate (diisocyanate) and polyol (or other reagent), which are released as volatiles during application. As discussed in the Polyurethane Foam section, polyisocyanates can cause irritation on contact with skin,[12], respiratory complaints, and are suspected carcinogens.[8] Polyisocyanates vary in volatility, with the less volatile claimed to be safer. However, if heated (eg - drying using a heater, applying on a hot day, placing the tin close to a heat source etc), these highly reactive reagents can reach concentrations in air capable of initiating an allergic reaction.[12]

Natural Varnishes

Natural varnishes are not as durable as synthetic varieties, but are safer and healthier.[9]

New and restored wood may not need a varnish, and pure lineed oil or beeswax can give a good, non toxic finish. Of the natural varnishes, shellac, a resin distilled from tropical trees has been recommended as an effective solvent, and organic paint suppliers offer shellac with linseed oil instead of solvents.[9]

Stone Flooring

Stone, such as granite, sandstone, limestone, slate, quartzite or marble are extremely durable, high quality flooring alternative.[25]

Production

Resource Depletion
Stone is a non-renewable resource and requires quarrying of often very attractive landscape areas. Clumsy and destructive quarrying in unsuitable sites such as national parks can be extremely damaging, particularly if the extracting company is allowed to remove too large a quantity of stone from a particular area. Limestone is a particular case in point in the UK[23,33] and marble extraction has caused appaling environmental damage in Italy[23] and Rajasthan, India. Conversely, derelict areas or uninteresting countryside can even be improved by careful restoration following well planned extraction.[23] If extracted using appropriate methods, stone is environmentally more acceptable than the alternatives in long-life situations in quality buildings.[25]

Energy Use
The energy cost of transporting stone is high, and therefore local sources should be used where possible rather than, for instance, Carrara marble, to avoid this hidden and unneccessary environmental cost.[25]

Use

Health
Some granite contains fairly high levels of uranium, and levels of radon, the radioactive gas given off during decay of uranium, may need to be checked in buildings where this stone is used extensively.[23]

Recycling/Reuse/Disposal
Granite is the most durable building stone found in the UK and can be reused almost indefinately. Slate can be recycled several times before it begins to weaken along the cleavage planes.[23] Sandstone comes in varying strengths, and limestone is generally the weakest indigenous building stone.

UK carpet manufacturers are increasingly joining GUT, the Association of Environmentally Friendly Carpets, a German initiative which is gaining support from most of the major western European manufacturers.[28]

15.5 Alternatives

15.5.1 Alternative Carpet Fibres

There are a number of natural fibre products available, all of which are renewable plant resources, require minimal processing and are biodegradable. These alternatives have not been popular, due to the practical concern that they are not easy to clean.[30] However, they are highly durable,[23] and often considered aesthetically more pleasing than their synthetic counterparts.

(a) Sisal

Produced from the sub-tropical *Agave sisalana* bush, sisal forms a very hardwearing floorcovering,[23] which is reputed to be soft enough even for bedroom floors.[45] While potentially a renewable resource, there are concerns regarding the use of intensive cultivation techniques to produce sisal, with resulting degradation of soils.[53] Sisal dust is also thought to be responsible for chronic respiratory problems in sisal factory workers.[54]

(b) Seagrass

A grass grown in sea water paddyfields which forms an almost impermiable floorcovering which is totally anti-static. Seagrass is naturally stain resistant, but the down-side of this is that it cannot be dyed.[45]
Seagrass forms part of a delicate inter-tidal zone ecosystem, essential to the proper functioning of neighbouring mangrove swamps and coral reef colonies. While the impact of seagrass harvesting is currently small compared to that of pollution and clearance for development,[54] it nevertheless places an additional stress on these fragile ecosystems.

(c) Coir

A hairy fibre beaten from coconut husk, woven to form carpet or carpet tiles.[45] Its most common usage is as door matting.[23]
Coir is a waste material, and its use should therefore be encouraged. However, there is concern over the environmental impacts of the 'retting' process. For example, in Kerala, India, this involves soaking the husks for up to 12 months, which is reported to result in anoxic conditions in the estuaries used for the process, causing extensive damage to aquatic life in the area.[56]

(d) Cotton

In order to be a useful flooring material, cotton is loop-woven to give body and 'pile'. While easier to clean than the other alternative fibres, cotton carpet is not good for heavy use areas and the pile tends to flatten with age.[45] Although *potentially* environmentally benign, cotton production can have severe ecological impacts, particularly when intensively cultivated with high chemical applications of pesticides and fertiliser. Irrigation demands can upset the local water balance, as happened around the Aral sea.[23]

Blends of natural fibres are also available, such as 'Sisool', a wool/sisal blend manufactured by Crucial Trading Ltd, which "combines the softness of wool with the durability of sisal".[45]

(e) Hessian

Hessian is produced from Jute fibres, which were first exported from India in the early 19th century, and is traditionally used for carpet and linoleum backing, rugs and carpet.[23] While not as resilient as sisal, Jute is somewhat softer.[45]

15.5.2 Smooth Floorings

(a) Exposed Floorboards

Exposed floorboards can be an attractive and low impact flooring, provided they are finished with water based varnish, wax or linseed oil, rather than synthetic varnish.

(b) Earth Floors

Traditional earth floors in asia and Africa are made from compacted stone or soil, and smoothed with a mixture of soil and cow dung, which gives resistance to abrasion, cracks and insects. Other surface hardeners include animal urine mixed with lime, ox blood mixed with cinders and crushed clinker, animal glues, vegetable oils, powdered termite hills and crushed shells.
Alternatively, an earth floor can be constructed by covering a subbase of well compacted clay-rich soil with large sized gravel (to stop capaillary action), topped by pea gravel and soil, the surface layer made from silty soil mixed with 5% linseed oil and compacted with tamper or vibrator. Earth floors are only really suitable for reasonably dry areas with good drainage and a low water table.[57]

1. Flexible Polyurethane Foam Glossary. Polyurethane Foam Association. http://www.usit.net/hp/pfa/glossary.html. 1995

2. Proving Chemically Induced Asthma Symptoms - Reactive Airways Dysfunction Syndrome, a new medical development. (R. Alexander) http/www.seamless.com/talt/txt/asthma/html. 1996

3. Reactive Airways Dysfunction Syndrome (RADS). (S.M. Brooks, M.A. Weiss & I.L. Bernstein). Chest, Vol. 88 (3) 376-384. 1985

4. Thermoplastic polyurethane 1. (Schilberg & Katie) http://eetsg 22.bd.psu.edu/research/projects/pu/pu.html. 1995

6. Indoor Chemistry: Ozone, Volatile Organic Compounds and Carpets. (C.J. Welschler, A.T. Hodgson & J.D. Wooley. Environmental Science and Technology. vol. 26 2371-2377. 1992

7. Carpet and Indoor Air Quality (The Carpet & Rug Institute) http://www.Dalton.net/cri/indoor.html . 1995

8. The Green Home Handbook. (G. Marlew & S. Silver) Fontana. 1991

9. H is for ecoHome. (A. Kruger) Gaia Books Ltd, London. 1991

10. The Non-toxic Home. (D.L. Dadd) Jeremy P. Tarcher Inc, Los Angeles. 1986

11. The Consumers Good Chemical Guide. (J. Emsley) W.H. Freeman & Co. Ltd, London. 1994

12. Buildings and Health - the Rosehaugh Guide to Design, Construction, Use and Management of Buildings (Curwell, March & Venables) RIBA Publications 1990.

13. Naughtly Nylon Creates a Hot and Bothered Atmosphere. (Pearce) New Scientist vol.129 p.24, March 16 1991

14. Hazardous Building Materials. (S.R. Curwell & C.G. March) E & F.N. Spon Ltd, London. 1986

15. Green Building Digest Issue no. 5, August 1995

16. C for Chemicals - Chemical Hazards and how to avoid them. (M. Birkin & B. Price) Green Print/Merlin Press Ltd, London. 1989

17. PVC: Toxic Waste in Disguise.(S. Leubscher (Ed).) Greenpeace International, Amsterdam. 1992

18. Achieving zero dioxin - an emergency stratagy for dioxin elimination. Greenpeace International, London. 1994

19. The World Environment 1972-1992. Two Decades of Challenge. (M.K. Tolba & O.A. El-Kholy (Eds)) Chapman & Hall, London, for the United Nations Environment Program. 1992

20. Production and Polymerisation of Organic Monomers. IPR 4/6. Her Majesties Inspectorate of Pollution, HMSO , London 1993

21. Environmental News Digest vol. 10 no.1, Friends of the Earth, Malaysia. Jan. 1992

22. Ecological Building Factpack. (R. Pocock & B. Gaylard) Tangent Design Books. 1992

23. Eco-renovation - the ecological home improvement guide. (E. Harland) Green Books Ltd 1993.

24. Green Design - A Guide to the Environmental Impact of Building Materials (A. Fox & R. Murrell) 1989

25. The Green Construction Handbook - A Manual for Clients and Construction Professionals. (Ove Arup & Partners) JT Design Build Publications, Cedar Press, Bristol. 1993

26. Dictionary of Building 4th Edition: (J.H. Maclean & J.S. Scott). Penguin Books Ltd, London 1993

27. Eco-Renovation - the Ecological Home Improvement Guide. (E. Harland) Green Books Ltd, Devon 1993

28. Keynote Market Report - Carpets and Floorcoverings (D. Jones, Ed). 1995

29. Retail Business No. 453 The Economist Intelligence Unit, London November 1995

30. RETAIL BUSINESS December 1995.. No. 454 The Economist Intelligence Unit, London

31. Sources of pollutants in indoor air.. (H.V. Wanner) In: B. Seifert, H.J. Van der Weil, B. Dodet & I.K. O' Niell (Eds). Environmental Carcinogens - Methods of Analysis and Exposure Measurements, Volume 12: Indoor Air. IARC Scientific Publications, no. 109. 1993.

32. Indoor Air Pollution: Problems and Priorities. (G.B. Leslie & F.W. Lanau) Cambridge University Press. 1992

33. Green Building Digest No. 1, January 1995

34. Environmental Building News. Vol.5 1., p10-12. Jan/Feb 1996

35. DWL Linoleum 1994 Catalogue

36. Khars Parquet & Inspiration International Idea Magazine - Floors. 1995.

37. ENDS Report 243, April 1995 p.23-24

38. ENDS Report 235, August 1994

39. The Interaction Between Dust, Micro-Organisms and the Quality of Cleaning. (A. Abildgaard) In: B. Berglund, & T. Londvall (Eds). Healthy Buildings. Swedish Council for Building Research, Stockholm, 3, 197-205. 1988

40. Ethical Consumer Issue 26, October/November 1993. ECRA Publishing Ltd.

41. Fashion Victims. The Globe, No.26.

42. Wool Record October 1994.

43. Scab Wars. Friends of the Earth 1993..

44. Green Building Digest No. 4, July 1995.

45. Crucial Trading 1995 Edition.

46. Ethical Consumer, Issue 33, Dec/Jan 1994.

47. Women, clothing and the Environment. Womens Environmental Network 1993.

48. ENDS Report 225, October 1993.

49. Ecollection by Esprit information, 1994.

50. Greenpeace Germany Recycling Report, 1992.

51. Spon's Architects and Builders Pricebook 1995.

52. US Consumer Safety Commision: Volatile Organic Emissions from Carpets, Report LBL31916. Lawrence Berkley Lab., California, 1992.

53. Nutrient Depletion of Ferralsols under hybrid sisal cultivation in Tanzania. A.E. Hartemink & A.J. Van Kekam. Soil Use and Management 10 (3) 103-107 1994.

54. 1994.. Respiratory Function and Immunological Reactions in Sisal Workers. Zuskin, Kanceljak, Mustajbegovic, Schachter & Kern. International Arch. Occupational Environmental Health, 66 1. 37-42 1994.

55. Marine Pollution Bulletin 23 p113-116 1991..

56. Pollution Indicators of coconut husk retting areas in the Kayels of Kerala. S.B. Nanden & P.K. Abdul-Aziz. International Journal of Environmental Studies, 47 1. p19-25 1995.

57. Appropriate Building Materials R. Schultz & K. Mukerji. Skat & IT Publications Ltd 1993.

16.1 Books & Factsheets

AD Profile 106. Contemporary Organic Architecture. 1993.

Allen, P and Todd, B. Off the Grid. Managing Independent Renewable Electricity Systems. *C.A.T. Publications* 1995, Price: £5.50

Anink, D., Boonstra, C. and Mak, J. Handbook of Sustainable Building: an Environmental Preference Method of Selection of Materials for Use in Construction and Development. *James & James* 1996.

Baggs, S & J. The Healthy House. *Thames and Hudson* 1996, London.

Borer P and Harris C. The Whole House Book. Ecological Building Design and Materials 1997. Available August from CAT

Borer, P and Harris, C, Out of the Woods, Ecological Designs for Timber-Frame Housing *W.S.T./C.A.T. Publications* 1994, Price £12.50

Boyle, G. Renewable Energy. *Centre for Alternative Technology* 1996, Price £22.50

Broome, J and Richardson, B. The Self Build Book. *Green Books*, 1991.

Crosbie Michael J. *Green Architecture* 1994 ISBN 1-56 496-153

Curwell, S.R. Fox, R.C and March, C.G. Use of CFC's in Buildings *Fernsheet Ltd* 1988

Curwell, S R and March C G, Hazardous Building Materials. A Guide to the Selection of Alternatives *E & FN Spon.* 1986

Day, C. Building with Heart. *Green Books* 1990. Price £9.95.

Day, C. Places of the Soul. Architecture and Environmental Design as a Healing Art. *The Aquarian Press.* 1990 Price £12.99

Edwards, B. Towards Sustainable Architecture. 1996

Farmer J. Green Shift. *Butterworth-Heineman* 1996.

Fox, A and Murrell, R. Green Design. A Guide to the Impact of Building Materials. *Architectue Design and Technology Press.* 1989

Franck, K A and Ahrentzen, S. New Households New Housing. *Van Nostrand Reinhold.* 1991

Goulding JR et al (ed) Energy Conscious Design 1993. A Primer for Architects. Batsford, Commission of EC EUR 13445.

Hail, K and Warm, P. Greener Building. Product and Services Directory. *The Green Building Press.* 1995.

Harland, E. Eco-Renovation. The ecological home improvement guide. *Green Books in assoication with the Ecology Building Society.* 1993. Price £9.95

Hawkes D. Environmental Tradition: Studies in the Architecture of the Environment. 1995.

Holdsworth and Sealey. Healthy Buildings. *Longman.* 1992.

Jackson, F. Save Energy, Save Money: A Guide to Energy Conservation in the Home. New Futures *C.A.T.* 1995 Price £4.50

Houben H and Guillaud H. Earth Construction. *Intermediate Technology Publication, London* 1994.

Johnson, J., and Newton, J. Building Green: A Guide to Using Plants on Roofs, Walls and Pavements. 1993 Illustrated £14.95

Johnson, S. Greener Buildings. -The environmental impact of property. *The MacMillan Press Ltd.* 1993.

Kahane, J. Local Materials. A self-builder's manual. *Quick Fox Publications.* 1978.

Lewis, Owen and Guilding, John (Eds). European Directory of Sustainable and Energy Efficient Building. *James & James* 1995. ISBN 1-873936-36-2.

Lewis O et al. Sustainable Building For Ireland 1996. *The Stationery Office, Dublin* (ISBN 0 7676 2392-8) EUR 16859-016761703

McCamant, K and Durrett, C. Cohousing. T*en Speed Press* 1989. Price £15.99.

Merchang, C. Radical Ecology. The Search for a Livable World. *Routledge* 1992. Price £11.99

Papanek, V. The Green Imperative. Ecology and Ethics in Design and Architecture. *Thames and Hudson,* 1995.

Papanek, V. Design for the Real World. Human Ecology and Social Change. *Thames and Hudson.* 1992 Price £10.95.

Pearson, D. The Natural House Book. *Conran Octopus* 1989.

Pearson, D. Earth to Spirit. In search of Natural Architecture. *Gaia Books.* 1994.

Pearson, D. The Natural House Catalog. Everything you need to create an environmentally friendly home. *Simon and Schuster Inc.* 1996. Price £19.99.

Pocock, R and Gaylard, B. Ecological Building Factpack. A DIY guide to energy conservation in the home and strategies for green building. *Tangent Design Books.* 1992. Price £4.95.

Potts, M. The Independent Home. Living Well with Power from the Sun, mind and Water. *Chelsea Green* 1993. Price £14.95.

Roodman, D. M. and Lenssen, N. A Building Revolution. How Ecology and Health Concerns are Transforming Construction. *The Worldmatch Institute* 1995.

Smith, L. Investigating Old Buildings. *Batsford Academic and Educational, London.* 1985. Price £19.95.

Solar Architecture in Europe (1991) *Prism Press (Bridport Dorset)* ISBN 1 85 327073 3 EUR 12738

Steen, A.S, Steen, B. and Bainbridge, D. The Straw Bale House. *Chelsea Green*. 1994.

Talbott, J. Simply Build Green. A Technical Guide to the ecological Houses at the Findhorn Foundation. *Findhorn Press*. 1995. Price £9.95.

Vale, B. and R. Green Architecture. Design for a sustainable future. *Thames and Hudson Ltd*. 1991.

Van der Ryn, S. The Toilet Papers. Recycling Waste and Conserving Water. *Ecological Design Press*. 1995.

Van Der Ryn, S. and Calthorpe, P. Ecological Design. *Island Press*. 1996.

Victor Papanek. The Green Imperative: Ecology and Ethics in Design and Architecture. 1995.

16.2 Magazines & Journals

Building for a Future. Magazine of the Association for Environment Conscious Builders. Nant-y-Garreg, Saron, Llandysul, Carmarthenshire SA44 5EJ. Tel. 01559 370908

Eco-Design. The magazine of the Ecological Design Association. EDA, The British School, Slade Road, Stroud, Gloucestershire, GL5 1QW. Tel 01453 765575

Environmental Building News. RR1, Box 161, Brattleboro, VT 05301 USA. Tel. + 802/257 7300

The Green Building Digest. Issues 13–16 not included in this publication, available from Department of Architecture, The Queen's University of Belfast, 2–4 Lennoxvale, Belfast BT9 5BY, Tel. 01232 335466, or ECRA Publishing, 41 Old Birley Street, Manchester M15 5RF. Tel. 0161 226 2929

The Journal of Sustainable Product Design. The Centre for Sustainable Design, The Surrey Institute of Art and Design, Falkner Road, Farnham, Surrey GU9 7DS. Tel. 01252 732229

16.3 Useful Organisations

ACTAC – The Technical Aid Network, c/o Eddie Walker, 25 Bexley Road, Harehills, Leeds LS8 5NS. Tel. 0113 2493491. *Former producers of the Green Building Digest. ACTAC is a charity dedicated to promoting sustainable neighbourhood development through community participation. Although its status as an organisation is in doubt, it is aiming to continue as a network for information exchange.*

Association for Environment Conscious Building (AECB), Nant-y-Garreg, Saron, Llandysul, Carmarthenshire SA44 5EJ. Tel. 01559 370908. *Produce a directory of environmentally preferable products, manufacturers and contractors and Building for a Future magazine.*

Centre for Urban Ecology, 318 Summer Lane, Birmingham, B19 3RL. Tel. 0121 359 7462

Department of Architecture, The Queen's University of Belfast, 2–4 Lennoxvale, Belfast BT9 5BY. Tel. 01232 335466. *Current producers of the Green Building Digest. Tom Woolley has established an office at Queen's University to carry out our research, design and development work in the field of green building.*

Ecological Design Association (EDA), The British School, Slad Road, Stroud, GL5 1QW. Tel. 01453 765575. *Produce EcoDesign magazine.*

Friends of the Earth, 26–28 Underwood St, London N1 7JQ. Tel. 0171 490 1555

Greenpeace, Cannonbury Villas, London, N1 2PN. Tel. 0171 865 8100

SALVO, Ford Woodhouse, Berwick upon Tweed, TD15 2QF. Tel. 01668 216494. *Produce a directory listing suppliers of reclaimed materials.*

The Centre for Alternative Technology, Machynlleth, Powys, SY20 9AZ. Tel. 01654 702400. *Produce a whole library of 'green' literature.*

The Ethical Consumer Research Association (ECRA), 41 Old Birley Street, Manchester M15 5RF. Tel. 0161 226 2929. *Produce the Ethical Consumer Magazine, which researches the environmental and ethical records of the companies behind the brand names, and promotes the ethical use of consumer power.*

The London Hazards Centre, Interchange Studios, Dalby Street, London, NW5 3NQ. Tel. 0171 267 3387

The Ethical Consumer Research Association

ECRA is a not-for-profit voluntary organisation managed by its staff as a workers' co-operative. It exists to promote universal human rights, environmental sustainability and animal welfare by encouraging an understanding of the ability of ethical purchasing to address these issues, and to promote the systematic consideration of ethical and environmental issues at all stages of the economic process.

Background

ECRA began life in June 1987 as a research group collecting information on company activities. In March 1989 it launched the Ethical Consumer magazine and attracted 5,000 subscribers by the end of its first year. Since then it has gone on to develop a range of other campaigns, products and services (see below).

In 1991 it raised £40,000 in loans from readers of the magazine. This investment has provided the main financial base for ECRA's activities. Since much of ECRA's work is not eligible for grant funding, we normally rely on the sale of publications and other information to pay for our operations.

EC Research

Besides their research into the environmental impact of building materials for the Green Building Handbook and Digest, ECRA operates a research service providing information on:

♦ Company activities and corporate responsibility issues
♦ Product environmental impact and lifecycle analysis
♦ Establishing ethical purchasing policies within institutions.

ECRA Publishing

As well as researching the Green Building Handbook and the Green Building Digest magazine ECRA also produces:

♦ The bi-monthly EC magazine and Research Supplement;
♦ The Corporate Critic on-line database. The database contains abstracts taken from other publications which are critical of specific corporate activities indexed under a range of social and environmental headings. Users pay modest charges to access the information;
♦ The Ethical Consumer Guide to Everyday Shopping (a 230 page paperback book);
♦ Postcards which are designed to help EC magazine readers to write to companies explaining the ethical issues which have influenced them to avoid or select a particular product.

EC Campaigns

ECRA has recently begun working on three main campaign areas arguing for changes in the law which would:
♦ Permit and encourage institutions such as UK pension funds and local authorities to invest and purchase ethically,
♦ Require companies to provide greater information on social and environmental issues
(a) in their annual reports, and
(b) on product labels.
♦ Permit advertising (especially on TV) by campaign groups and companies which address consumer product-related social and environmental issues.

Institute for Ethical Purchasing Research

ECRA is currently seeking to establish an independent research organisation for those aspects of its research which are charitable. These are likely to include an internet database for consumers on product environmental issues, a publication on best practice by organisations which purchase ethically and a range of shorter reports on specific corporate responsibility issues.

The Association for Environment Conscious Builders

The Association for Environment Conscious Building (AECB) was established in 1989 and is a non profit making association which exists to increase awareness within the construction industry, of the need to respect, protect, preserve and enhance the environment, locally and globally. These principles have become to be known collectively as "sustainability".

The objective of the AECB is to facilitate environmentally responsible practices within building. Specifically the AECB aims to:-

♦ Promote the use of products and materials which are safe, healthy and sustainable
♦ Encourage projects that respect, protect and enhance the environment
♦ Make available comprehensive information and guidance about products, methods and projects
♦ Support the interests and endeavours of its members in achieving these aims

Designing, managing and carrying out renovations or new-build projects can be painstakingly difficult work. With the added burden of needing to ensure that discussions and choices are ecologically balanced, the task really becomes a mean feat.

This is where groups like the AECB really come into play. The AECB membership (architects, building professionals, energy consultants, electricians, suppliers, manufacturers, local authorities, housing associations etc etc) are all regularly receiving the latest information and research on the environmental aspects of construction as well as staying on top of their own professional service commitments. Construction professionals that have taken the time to study the ecological pros and cons of their industry are more likely to provide an all round quality service than those who have not.

The AECB helps and encourages its members to develop a documented environmental policy against which they can be measured for environmental performance. This move provides further momentum in the drive towards sustainability in the built environment and helps to brush aside the lip service attitude that can easily emerge in the environmental arena. AECB members are listed in a Year Book which is widely circulated.

Being 'green' is something which requires constant attention and there is rarely an easy answer. Every action has some sort of impact upon the environment. The AECB exists to try and lessen that impact. The AECB recognises that 'green' awareness can only be brought about through education and information. To this end the Association publishes information and participates in exhibitions, attends conferences and gives talks on subjects relating to 'greener' building. The AECB publishes information on the subject of ecologically sound building. Its magazine Building for a Future is produced quarterly. For assistance in searching and specifying green products, it produces a products and services directory, 'Greener Building' and 'GreenPro', construction research software. Whilst AECB membership is available to all, only those with an actual environmental policy are included in 'GreenPro' and 'Greener Building'. Entries are free for those listings that meet the necessary criteria. GreenPro 97 is available on discs or CD.

Green projects undertaken by any AECB member are eligible for the eco-certificate - SPEC (Sustainable Projects Endorsement Scheme). This scheme addresses aspects such as appropriateness; site issues; materials; energy; water and waste. It looks in detail at approaches to management, land use and enhancement of the site's ecology, embodied energy of materials, energy conservation measures and the handling of waste. The system is flexible enough to be applied not only to building work and landscaping but equally to modest jobs such as decorating or re-wiring.

For an information pack (sample issue of magazine, current Year Book and membership/subscription details), please send an A4 SAE (64p) to AECB, Nant-y-Garreg, Saron, Llandysul, Carms SA44 5EJ. Tel 01559 370908. You can also find out more about this unique organisation by looking at its Web Site which can be located at http://members.aol.com/buildgreen/